AF903319

At War with the Single Strand

Renu Swarup
Govindarajan Padmanaban
Narendra Kumar Arora
Editors

At War with the Single Strand

How Indian Science Fought the COVID-19 Pandemic

Springer

Editors
Renu Swarup
Department of Biotechnology
Former Secretary to Government of India,
Ministry of Science and Technology
New Delhi, Delhi, India

Govindarajan Padmanaban
Department of Biochemistry
NASI Honorary Scientist, Former Director,
Indian Institute of Science
Bengaluru, Karnataka, India

Narendra Kumar Arora
The INCLEN Trust International
Executive Director
New Delhi, Delhi, India

ISBN 978-981-96-8121-1 ISBN 978-981-96-8122-8 (eBook)
https://doi.org/10.1007/978-981-96-8122-8

© The Editor(s) (if applicable) and The Author(s), under exclusive license to Springer Nature Singapore Pte Ltd. 2025

This work is subject to copyright. All rights are solely and exclusively licensed by the Publisher, whether the whole or part of the material is concerned, specifically the rights of translation, reprinting, reuse of illustrations, recitation, broadcasting, reproduction on microfilms or in any other physical way, and transmission or information storage and retrieval, electronic adaptation, computer software, or by similar or dissimilar methodology now known or hereafter developed.
The use of general descriptive names, registered names, trademarks, service marks, etc. in this publication does not imply, even in the absence of a specific statement, that such names are exempt from the relevant protective laws and regulations and therefore free for general use.
The publisher, the authors and the editors are safe to assume that the advice and information in this book are believed to be true and accurate at the date of publication. Neither the publisher nor the authors or the editors give a warranty, expressed or implied, with respect to the material contained herein or for any errors or omissions that may have been made. The publisher remains neutral with regard to jurisdictional claims in published maps and institutional affiliations.

This Springer imprint is published by the registered company Springer Nature Singapore Pte Ltd.
The registered company address is: 152 Beach Road, #21-01/04 Gateway East, Singapore 189721, Singapore

If disposing of this product, please recycle the paper.

Foreword

डॉ. विनोद कुमार पॉल
सदस्य
Dr. Vinod K. Paul
MEMBER

सत्यमेव जयते

भारत सरकार
नीति आयोग, संसद मार्ग
नई दिल्ली–110 001
Government of India
NATIONAL INSTITUTION FOR TRANSFORMING INDIA
NITI Aayog, Parliament Street
New Delhi-110 001
Tele. : 23096809 Telefax : 23096810
E-mail : vinodk.paul@gov.in

19 April, 2024

FOREWORD

COVID 19 was the biggest public health crisis faced by humanity in centuries. The world also witnessed social and economic challenges bringing normal life to a standstill. The most effective tool to manage this public health crisis was Science. Four years now since the Pandemic, nationally and globally there is an effort to document the key learnings and reflect on what went well, where we could do better and what is required for the future. The efforts made by India to manage this crisis have been very well recognised and lauded globally. Our Indian science fraternity contributed so efficiently to this unusual war, working hand in hand with the public health team and others. The whole-of-Government approach was the key determinant of sucess.

The Book **'At war with the single strand : How the Indian Science fought the COVID 19 Pandemic'** has documented not just the success stories, but also the challenges faced in the scientific management of the pandemic. The editors and the contributors have brought out their personal hands-on experience in each of the issues. From the details of the science of the Pandemic to the science behind the vaccines; from the public health strategy on test, track and treat, to the make-in-india diagnostic success story; from the indigenous vaccine development and manufacture, to the world's largest adult vaccination programme using the digital platform CoWin; and from the details of the robust research, translation and clinical trial network, to the road map ahead, this publication is a very comprehensive, well-articulated documentation of the India's science effort in fighting the war against the virus.

In this fight, we saw how Governance, Leadership and Collaboration played a major role in the success. The Empowered Groups set up by the Government were entrusted the responsibility of addressing the challenges and finding immediate solutions. This facilitated quick decision-making which helped the country in managing the pandemic and also built a strong ecosystem for the future.

I would like to compliment the Editors — Dr Renu Swarup, Prof G Padmanabha and Dr N K Arora. They have edited the book with great depth, passion and got the best minds and best experts in each of the areas to give a detailed account of their contribution to the management of the pandemic. The contributors have been very carefully chosen to cover the entire landscape. I am hopeful that this book will be a very useful reading and reference material for all scientists, researchers, policy makers who will be able to pick up the key points during future planning and implementation of the national epidemics.

This book will also be an interesting reading material for the young researchers as well. They will be motivated to take up scientific challenges. The book will also give common citizen also give them a detailed understanding of the scientific issues in plain easy-to-understand language. This will also be an important document for the International agencies to get a first-hand account of how India fought this war so effectively.

I once again congratulate and wish the editors and contributors all the very best for the success of this book.

(Vinod Paul)

स्वच्छ भारत
एक कदम स्वच्छता की ओर

Preface: India's Scientific response to COVID -19

The single stranded RNA virus brought the world to its knees. It was a war of humanity with the tiny virus. All our strategies and action plans were put to test to make sure that the single stranded RNA virus does not outsmart us. We were caught unawares. Science was our biggest strength, all the knowledge, tools, platforms, technologies, and products were brought forward to fight this war. We had a huge dependency on imports, and with supply chain disruptions and lockdowns, there was a need not just for new innovation in research but also innovation in models of operation.

This publication is an enumeration of the Journey through these difficult times and how Science in India played a central role in fighting this war against COVID.

If we reflect on the year 2020 there are two things that immediately come to our mind—the pandemic—the COVID, the challenge that came to us suddenly in the beginning of the year 2020, and second is how Science came forward to respond to this challenge. So while it was a year full of challenges it was also a year of opportunities which brought forward the strength of Science to help us as a nation, as a global community to fight COVID. With the onset of the Pandemic in early 2020, the world focused its attention on the Scientists and researchers for finding science-based solutions to fight this war.

A tiny RNA virus had brought the world to a grinding halt and impacted the health, economy, and livelihood of mankind. India's Science & Technology-based response to the Pandemic has been phenomenal and has given our Country an elated level of confidence to face future challenges. The successful tools and technologies developed by the scientists in the country to fight COVID made the country not only self-reliant but also in a position to respond to the Global needs .Today we have the required competence and capacity to use new cutting-edge technologies to develop solutions for tackling other diseases. This was made possible through a very strong industry academia collaboration engaging a network of laboratories involving public and private sector and also start-ups.

We proudly completed the largest vaccination program of over two billion vaccine doses. India not only made vaccines available for its own population but also supplied vaccines to a large number of countries. From vaccine development and manufacture to the largest immunization program, which was facilitated by the Indigenously developed Digital platform CoWin, this publication gives a first-hand account of these amazing feats. It documents the stories from the key players and brings out not only the success but also the challenges and how innovative solutions were found in the shortest time frame to address these.

India's Science & Technology-based response to the Pandemic has been phenomenal and has given our country an elevated level of confidence to face future challenges. The successful tools and technologies developed by the scientists in the country to fight COVID have made the country not only self-reliant but also in a position to respond to the Global needs. Today we have the required competence and capacity to use new cutting edge technologies to develop solutions for tackling other diseases.

The country also saw one of its largest vaccine development programs—Mission COVID Suraksha—the country developed the largest portfolio of vaccine candidates on different platform technologies—inactivated virus, mRNA, DNA, protein subunit, and other new platform technologies. India was known to be the world's Vaccine manufacturer, and it has now been globally recognized as a strength in vaccine development as well. This was made possible through a very strong industry academia collaboration engaging a network of laboratories involving public and private sector and also start-ups. An enabling robust ecosystem was built which can now be used for further vaccine developmental research. This knowledge, experience, and capacity is critical to accelerate the development of vaccines for priority diseases like Pan Corona virus, TB, HIV, Malaria, Chikungunya, Zika, and many others.

We have very proudly completed the largest vaccination program of over two billion vaccine doses. The digital platform CoWin was a key factor to facilitate this initiative. India has not only made vaccines available for its own population but also supplied vaccines to a large number of countries and also under the Access to Covid Tools (ACT) Accelerator.

Our Diagnostic success story makes each Indian proud, a true example of Atma Nirbhar Bharat. From complete imports in March 2020 to 100% self-reliance on indigenous kits in June 2020 to meet the growing needs of testing, our industry, academia, and startups came forward to work collaboratively. Today the country has the competence to develop the most sensitive and affordable kits which are available for the domestic and export market.

The rapidly evolving virus requires constant monitoring to understand the sequence changes and correlate the mutations clinically to have complete information on immune escape, disease severity, transmissibility, etc. The country's largest Genome sequencing network, Indian SARS-CoV-2 Consortium on Genomics (INSACOG), was set up. This became a major contributor from India to the Global genomic sequence database. This consortia along with the state sentinel sites and hospital network continuously tracked the virus and samples from across the country and international travelers were sequenced daily. It gave us the ability to detect new variants and correlate clinically.

SARS-CoV-2 is not the last virus we are fighting a war against, with the evolving virus and its numerous mutations there will be a continued focus on science based solutions to address the challenges of human health. This ecosystem which we have built was put to test during the pandemic with high levels of success. Many other healthcare priorities also need urgent science-based solutions which this strong ecosystem is now ready to deliver. The collective fight against this pandemic was not

Preface: India's Scientific response to COVID -19

The single stranded RNA virus brought the world to its knees. It was a war of humanity with the tiny virus. All our strategies and action plans were put to test to make sure that the single stranded RNA virus does not outsmart us. We were caught unawares. Science was our biggest strength, all the knowledge, tools, platforms, technologies, and products were brought forward to fight this war. We had a huge dependency on imports, and with supply chain disruptions and lockdowns, there was a need not just for new innovation in research but also innovation in models of operation.

This publication is an enumeration of the Journey through these difficult times and how Science in India played a central role in fighting this war against COVID.

If we reflect on the year 2020 there are two things that immediately come to our mind—the pandemic—the COVID, the challenge that came to us suddenly in the beginning of the year 2020, and second is how Science came forward to respond to this challenge. So while it was a year full of challenges it was also a year of opportunities which brought forward the strength of Science to help us as a nation, as a global community to fight COVID. With the onset of the Pandemic in early 2020, the world focused its attention on the Scientists and researchers for finding science-based solutions to fight this war.

A tiny RNA virus had brought the world to a grinding halt and impacted the health, economy, and livelihood of mankind. India's Science & Technology-based response to the Pandemic has been phenomenal and has given our Country an elated level of confidence to face future challenges. The successful tools and technologies developed by the scientists in the country to fight COVID made the country not only self-reliant but also in a position to respond to the Global needs .Today we have the required competence and capacity to use new cutting-edge technologies to develop solutions for tackling other diseases. This was made possible through a very strong industry academia collaboration engaging a network of laboratories involving public and private sector and also start-ups.

We proudly completed the largest vaccination program of over two billion vaccine doses. India not only made vaccines available for its own population but also supplied vaccines to a large number of countries. From vaccine development and manufacture to the largest immunization program, which was facilitated by the Indigenously developed Digital platform CoWin, this publication gives a first-hand account of these amazing feats. It documents the stories from the key players and brings out not only the success but also the challenges and how innovative solutions were found in the shortest time frame to address these.

From understanding the nature of the virus, to the Science behind the Vaccine, Experts have unraveled these complex issues and explained how Science plays a significant role in developing immediate solutions for urgent problems. The book gives a detailed account of the Robust Research and translation ecosystem which was built and strengthened in response to the Pandemic for developing medical countermeasures, including the diagnostic testing framework, the hospital network, clinical trial sites, shared facilities, and a Rapid Response regulatory mechanism.

Our diagnostic success story makes each Indian proud, a true example of Atma Nirbhar Bharat. From complete imports in March 2020 to 100% self-reliance of indigenous kits in June 2020. Today the country has the Competence to develop the most sensitive and affordable kits which are available for the domestic and export market.

The rapidly evolving virus requires constant monitoring to understand the sequence changes and correlate the mutations clinically to have complete information on immune escape, disease severity, transmissibility, etc. The country's largest Genome sequencing network, Indian SARS-CoV-2 Consortium on Genomics (INSACOG) was set up. This became a major contributor to the global genomic sequence database. This consortia along with the state sentinel sites and hospital network continuously tracked the virus and samples from across the country and international travelers were sequenced daily. This is now ready for expansion for surveillance of other diseases.

Based on these learnings both national and global, the road map ahead has also been elucidated.

This publication is a comprehensive documentation of how the virus and disease evolved in India and how Science fought the Pandemic and contained the virus and its spread. This book starts appropriately with a Foreword by Dr. Vinod Paul, Member, NITI Aayog, followed by a Comprehensive Executive Summary by the Editors. Section 2 starts with an enumeration of the Science behind the Pandemics, genome sequence-based understanding of the virus evolution and setting up of INSACOG, testing and treating in clinical settings, success story behind development of indigenous Diagnostic kits, completely replacing imports. This is followed by a vivid description of the true story behind development of indigenous vaccines and clinical trial network. The last segment deals with academia-industry partnership, Start-up response, Digital India to the fore, and our global learnings.

Each Contributor has played a very important role in this fight and they have in their own words conveyed their experience, the success, and challenges.

We hope the readers enjoy reading this book and get a good understanding of the Science that was behind the management of COVID-19. We also hope that this will be a documentation for posterity and will serve as a good reference on how Science responds to a public health crisis.

Once again Science has exhibited its strength in finding solutions for societal challenges.

New Delhi, India — Renu Swarup
Bangalore, India — Govindarajan Padmanaban
New Delhi, India — Narendra Kumar Arora

Acknowledgment

We, the Editors, would like to express our heartfelt gratitude to Dr. V. K. Paul, Member (Health) NITI Aayog, for his support and guidance in India's fight against COVID-19. His visionary leadership and deep commitment to public health have provided both strategic direction and inspiration throughout the COVID journey. We thank him for writing the Foreword for this book.

We would like to sincerely thank all the chapter authors and co-authors, whose scholarly inputs and dedication have been the backbone of this publication. Each contributor has played a critical role in our fight against COVID. Each chapter is a result of their hard work, in-depth knowledge, and passion for advancing health research and policy. We recognize the efforts of their respective teams as well, who worked tirelessly behind the scenes to ensure that accurate, insightful information is presented to the readers.

A special thanks to Ms. Renu Shrivas, Research Officer, INCLEN and the INCLEN Support team, for all the effort of coordination with all the authors and for maintaining consistent communication with the publishers and authors.

We are thankful to each individual and team member whose efforts have helped shape this publication.

Renu Swarup
Govindarajan Padmanaban
Narendra Kumar Arora

Executive Summary

At War with the Single Strand: How Indian Science Fought the COVID-19 Pandemic—This publication documents the journey of how the Indian Science experts and organizations fought the Pandemic. This publication is an edited compilation with contributions from more than 15 researchers, industries, experts, and policymakers who have played a major role in our fight against the virus. The publication enumerates the details of each of these success stories and also the challenges faced. The development of a robust ecosystem, development of innovative, affordable medical counter measures, and implementation of a well-planned strategy for countering and managing the pandemic have been detailed in this publication directly in the words of the players who were responsible for each of these activities.

If we reflect on the year 2020, there are two things that immediately come to our mind—the pandemic—the COVID, the challenge that came to us suddenly in the beginning of the year 2020, and secondly the Science that came forward to respond to this challenge. So while it was a year full of challenges it was also a year of opportunities for Scientific solutions to help us as a nation, as a global community to fight COVID. With the onset of the Pandemic in early 2020, the world focused its attention on the scientists and researchers for finding science based solutions to fight this war. A tiny RNA virus brought the world to a grinding halt and impacted the health, economy, and livelihood of mankind. The focus was on Science & Technology to find affordable solutions for understanding and tackling the virus. From Genome sequencing to Diagnostics to Vaccines and antivirals, Biotechnology has been the key sector which very efficiently delivered new technologies and products to address the key challenges of COVID.

It is important to look back to see how we began—it was late 2019 December when we were all suddenly informed about this SARS-CoV-2 strain and then as we moved into January, we began the discussion on what would be the possible options that science will have to put forward to fight this Pandemic.

Our R&D Strategy was prepared focusing on four major aspects:

(i) Understanding the virus—Genome sequencing
(ii) Testing—Diagnostics, specifically molecular diagnostics
(iii) Treatment—Novel therapeutics, natural products, phytopharma
(iv) Prevention—Development of novel vaccines

India's Science & Technology-based response to the Pandemic has been phenomenal and has given our country an elevated level of confidence to face future challenges. The successful tools and technologies developed by the scientists in the country to fight COVID have made the country not only self-reliant but also in a position to respond to the Global needs. Today we have the required competence and capacity to use new cutting edge technologies to develop solutions for tackling other diseases.

The country also saw one of its largest vaccine development programs—Mission COVID Suraksha—the country developed the largest portfolio of vaccine candidates on different platform technologies—inactivated virus, mRNA, DNA, protein subunit, and other new platform technologies. India was known to be the world's Vaccine manufacturer, and it has now been globally recognized as a strength in vaccine development as well. This was made possible through a very strong industry academia collaboration engaging a network of laboratories involving public and private sector and also start-ups. An enabling robust ecosystem was built which can now be used for further vaccine developmental research. This knowledge, experience, and capacity is critical to accelerate the development of vaccines for priority diseases like Pan Corona virus, TB, HIV, Malaria, Chikungunya, Zika, and many others.

We have very proudly completed the largest vaccination program of over two billion vaccine doses. The digital platform CoWin was a key factor to facilitate this initiative. India has not only made vaccines available for its own population but also supplied vaccines to a large number of countries and also under the Access to Covid Tools (ACT) Accelerator.

Our Diagnostic success story makes each Indian proud, a true example of Atma Nirbhar Bharat. From complete imports in March 2020 to 100% self-reliance on indigenous kits in June 2020 to meet the growing needs of testing, our industry, academia, and startups came forward to work collaboratively. Today the country has the competence to develop the most sensitive and affordable kits which are available for the domestic and export market.

The rapidly evolving virus requires constant monitoring to understand the sequence changes and correlate the mutations clinically to have complete information on immune escape, disease severity, transmissibility, etc. The country's largest Genome sequencing network, Indian SARS-CoV-2 Consortium on Genomics (INSACOG), was set up. This became a major contributor from India to the Global genomic sequence database. This consortia along with the state sentinel sites and hospital network continuously tracked the virus and samples from across the country and international travelers were sequenced daily. It gave us the ability to detect new variants and correlate clinically.

SARS-CoV-2 is not the last virus we are fighting a war against, with the evolving virus and its numerous mutations there will be a continued focus on science based solutions to address the challenges of human health. This ecosystem which we have built was put to test during the pandemic with high levels of success. Many other healthcare priorities also need urgent science-based solutions which this strong ecosystem is now ready to deliver. The collective fight against this pandemic was not just of Science, but all together—the healthcare workers, the central and state government machinery, and each citizen of the country contributed to the overall effort in India. There was a very strong support from the top leadership, and a whole of Government approach was followed. A strong inter-departmental, interagency, center–state, and industry academia partnership was seen.

This publication documents the stories from the key players and brings out not only the success but also the challenges and how innovative solutions were found in the shortest time frame to address these challenges.

The publication also lists out the key areas of concern which need to be strengthened for us to be prepared for future pandemics and public health emergencies. The learnings from COVID are important and give us a good understanding of the Strategy and Road map ahead.

There are Research priorities and ecosystem components which we need to focus on even in times when we are preparing for future pandemics. A global lacuna to deal with SARS-2 virus is the absence of specific antiviral molecules to treat the disease. Most of them used were anti-HIV drugs. One also hopes that with the infrastructure now available to tackle the virus and the traditional strength on R&D in the area of Pharmaceuticals, India would also be able to develop specific antiviral drugs to kill the SARS-2 virus and treat the disease. The ultimate goal has to be elimination of the virus, which is a challenge.

The Foreword is written by Dr. Vinod Paul, Member, NITI Aayog, who was at the helm of affairs and chaired the Covid Task Force, the NEGVAC, the Empowered group on Emergency Management that played a critical role in spearheading the coordination between ministries agencies, organizations, academia, industry, and international bodies.

In the second there are 12 chapters, each focusing on an important component of the Scientific war against the pandemic.

Chapter 1 on The Science of the Pandemic unravels the complexities associated with this virus and focuses on the Evolution of the SARS-CoV-2 Virus Spike Glycoprotein during the COVID-19 Pandemic. This is important not just in the context of the COVID-19 pandemic but also to help prepare for future pandemics and develop new solutions.

Chapter 2 gives a detailed account of how a robust ecosystem was built for managing the virus and developing strong medical countermeasures. From investments to capacities—both Infrastructure and Human resource and fostering partnerships between academia, industry, start-ups, and international groups, there is a detailed account of this effort. The actions taken for strengthening and building a translational ecosystem from discovery to research to product development and manufacture is a key learning which allows us to now be prepared for future pandemics

Chapter 3 is a very critical component of this journey. Understanding the virus and tracking its evolution was key to developing clinical management strategies. The INSACOG played an important role in Genomic surveillance and identifying the evolving mutations of the virus. This sequence information helped develop diagnostics, vaccines, and clinical management strategies and is an important surveillance network for many other infectious pathogens of the future. The key experts who coordinated this have given a very lucid account of this crucial component.

Chapter 4 focuses on how the Scientific strength is leveraged to test, track, and treat. A public health perspective of this issue has been given in detail. The chapter gives a detailed account of the governance and management plans which played an important role in successful management of COVID-19.

Chapter 5 covers an important component of the COVID management and that is our Diagnostic success story. Three key players—CCAMP, AMTZ, and Mylab, have given a first-hand account of how the country moved in a short period of time from being 100% dependent on imports for our test kits to becoming 100% self-reliant and developing indigenous kits for the large number of tests which were conducted specially at the peak time like the delta wave. The manner in which forward and backward linkages were established to ensure that we have no dependency on any external supply chain is an important aspect which today has brought many key players to a level that they are not only recognized as national manufacturers but also as key players in the global ecosystem.

Chapter 6 is the key chapter on the Make in India Vaccines and covers the success stories of Bharat Biotech International Ltd, Serum Institute of India Pvt Ltd, Zydus Cadila Healthcare Ltd, Biological E Ltd, Gennova Biopharmaceuticals Ltd, Hester Biosciences Ltd, Indian Immunologicals Ltd. Each vaccine manufacturer has in his own words listed out the complete story of how each vaccine was developed and manufactured in a short period of time to be able to make indigenous supplies available for the world's largest immunization program. The success and challenges in developing novel platform technologies will now pave the way for developing vaccines for many other public health emergencies.

Chapter 7 gives us a very detailed account of what is the Science behind the development of these vaccines. Each platform technology has been discussed in detail. The chapter highlights what are the key learnings, and how these platform technologies will now be important for new vaccine development.

Chapter 8 gives a detailed account of the vast clinical trial network that was established across the country to facilitate a very robust trial of all vaccines which were developed. This network was critical in getting the data which was important to get regulatory clearances. This network will now be a strong asset for future vaccine and drug development initiatives.

Chapter 9 focuses on how the COVID pandemic scientific product development saw a very strong Industry Academia network which facilitated the research, translation, and product development. This partnership has built a strong trust and credibility between partners and is going to be very important going forward for future research.

Chapter 10 outlines a very key component of our fight against COVID, which was the *Start-up* Response. We saw a number of start-ups coming forward and developing important products ranging from Sanitizers, to PPEs, to Ventilators, patient monitoring devices, devices for ICUs, etc. The investment opportunities and the role of incubators and mentors have been highlighted. This documentation is important because it lists out the strength of the start-up ecosystem of our country and, based on the learnings and challenges, also allows us to develop policies and initiatives which can further strengthen this ecosystem.

Chapter 11 gives us a detailed understanding of one of the most successful digital tool developed indigenously—CoWin which was a crucial component for one of the most successful vaccine immunization program. This platform today gives us the confidence to launch more such adult vaccination programs.

Chapter 12 in the end gives us a very detailed understanding of the COVID-19 learnings and global experience. This is very important for us to develop our strategy and road map for preparing for any such future emergency.

In the end of the publication, this section summarizes the key aspects highlighted by each contributor and, based on their learnings and experience, lays a clear road map for the future.

This publication is a comprehensive documentation of how the virus and disease evolved in India and how Science fought the Pandemic and contained the virus and its spread. The stories narrated by the key players are written in a simple manner to take the scientific aspects to the readers in an easy-to-understand manner, and we hope this will make an interesting read for all.

Contents

About the Editors

Renu Swarup completed her PhD in genetics and plant breeding from Forest Research Institute, Dehradun, India, and post-doctoral research from the John Innes Centre, Norwich, UK. She is the Former Secretary of the Department of Biotechnology (DBT), Government of India; and Former Chairperson, of the Biotechnology Industry Research Assistance Council (BIRAC); she also held additional charge as the Secretary at the Department of Science and Technology (DST). Currently, she is the vice chair of the International Centre for Antimicrobial Resistance Solutions (ICARS), Copenhagen, Denmark, and a member of the National Start-up Advisory Council, Government of India, India. She is also the scientific expert on the Advisory board of international pandemic preparedness group for the Government of UK, and the Chair of the Government of India expert group on preparedness for future pandemics. Her involvement has been with issues related to policy planning and implementation. She was engaged in the formulation of the National Biotechnology Vision and Strategy in 2001, 2007, 2015, and 2021. As the Secretary of the Government of India, Department of Biotechnology, she led a network of 16 autonomous research institutes, two public sector undertaking (PSUs), and a research and development network of more than 5000 projects across more than 100 research institutes, universities, and laboratories in India. She also led the COVID-19 vaccine, diagnostic, and genome sequencing mission. She is a fellow of the National Academy of Sciences (NASI) India, a life member of the Trust for Advancement of Agricultural Sciences (TAAS), and a member of the Organization for Women in Science for the Develop World (OWSD).

Govindarajan Padmanaban completed his PhD at the Indian Institute of Science, Bangalore, India, and is the Former Director of the Indian Institute of Science Bangalore, India. Currently, he chairs several technical committees for government science and technology establishments and is a senior science innovation adviser of the Biotechnology Research Assistance Council (BIRAC) of the Department of Biotechnology (DBT), Government of India. His research interests include recombinant DNA research, liver cytochrome-P450 gene expression, understanding the multifarious roles of the Heme molecule, and biology of the malaria parasite. His research group discovered and characterized the heme-biosynthetic pathway and established its role in parasite development during the transmission stages and parasite virulence in its asexual stage of development. He has published over 180 papers in peer-reviewed national and international journals. He was also the co-chair of the COVID-19 vaccine evaluation committee of the Department of Biotechnology (DBT) in India.

Narendra Kumar Arora completed his MD in Pediatrics from the All India Institute of Medical Sciences (AIIMS), New Delhi, India, and MMSc in clinical epidemiology and biostatistics from the University of Newcastle, Australia. He is a member of numerous national scientific, research, and academic committees constituted under the Ministry of Health and Family Welfare (MoHFW), Department of Biotechnology (DBT), Biotechnology Research Assistance Council (BIRAC), Department of Science and Technology (DST), Indian Council of Medical Research (ICMR), and University Grants Commission (UGC) for maternal and child health in India. Currently, he is the chair of the COVID-19 working group of the National Technical Advisory Group on Immunization (NTAGI), India, and a chair of various COVID-19 vaccines, drugs, and diagnostic activities in India; he is a member of the WHO's Global Advisory Committee on Vaccine Safety (GACVS), and WHO's scientific and technical advisory group for maternal, new-born, child, and adolescent health and nutrition (STAGE); co-chair of STAGE working group on anemia. He has published several papers in peer-reviewed national and international journals.

Evolution of the SARS-CoV-2 Virus Spike Glycoprotein During the Covid-19 Pandemic

1

N. V. Joshi, M. Vijayasarathy, and P. Balaram

The coronavirus, like other viruses, contains all the information necessary for replication, multiplication, and propagation encoded in its genome. Peter Medawar, the English immunologist and writer, aptly defined a virus "*as a piece of bad news wrapped up in protein*". The Covid-19 pandemic has been driven by the coronavirus, SARS-CoV-2, which emerged from Wuhan, China, in late 2019, and quickly spread all over the world. This spherical virus, a complex, lifeless chemical entity, springs to life and multiplies once it infects a host cell and subverts the cellular biochemical machinery to serve its own ends of replication and propagation. The coronavirus may have emerged from an animal host. However, in crossing the zoonotic barrier while jumping to humans, it faces many challenges in adapting to an unfamiliar and often hostile host. Finding an appropriate receptor, invariably a host protein on the surface of human cells, is its first task. Its next step will be to use this bridgehead on the cell surface to penetrate the defences of the host cell membrane, forcing entry to deliver its cargo of genetic material, ribonucleic acid (RNA), in the case of the pandemic virus. The spike protein, which dots the surface of the spherical viral particle, is the Trojan Horse, welcomed by a cell surface receptor on human cells and aided by host proteins in breaching the membrane barrier. The virus encounters challenges on its way to finding its host cell, most frequently antibodies, a key component of the human immune response to a foreign microbial invader,

N. V. Joshi
Energy and Wetlands Research Group, Centre for Ecological Sciences, Indian Institute of Science, Bangalore, India

M. Vijayasarathy
Jawaharlal Nehru Centre for Advanced Scientific Research, Jakkur, Bangalore, India

P. Balaram (✉)
National Centre for Biological Sciences, Tata Institute of Fundamental Research, Bangalore, India
e-mail: pb@iisc.ac.in

© The Editor(s) (if applicable) and The Author(s), under exclusive license to Springer Nature Singapore Pte Ltd. 2025
R. Swarup et al. (eds.), *At War with the Single Strand*,
https://doi.org/10.1007/978-981-96-8122-8_1

which can intercept the virus before it reaches its destination. The nature of this immune response may be conditioned by past encounters with similar viruses. This immune interception is, of course, mediated by very specific chemical interactions that permit the antibody to recognize the foreign virus. Molecular recognition in biology is often characterized by exquisite specificity, a feature which led the famous German chemist, Emil Fischer, to invoke the metaphor of "locks and keys," well over a century ago.

The virus is a formidable foe. Taking advantage of the imperfect translation of the genetic information in the rapidly replicating virus, viral variants or mutants, which vary the chemistry of the spike protein are generated. Effectively, the variants are now disguised, evading recognition by host antibodies; immune escape mutants then reach the host cell receptor. Mutations can also enhance the binding of the spike protein to its receptor or facilitate the events that lead to breach of the host cell membrane. Mutations thus enhance the fitness of the virus. These variants are then naturally selected, a process well understood in the Darwinian view of biology.

The term coronavirus entered the scientific literature in 1968 in a brief editor's note in the journal *Nature*, which informed readers that a group of virologists had proposed the name for a new class of viruses that had been discovered as causative agents for human respiratory infections [1]. The name was derived from the corona-like halo seen around the virus, when it was first visualized under an electron microscope in 1967 [2]. The authors of the 1967 report, David Tyrrell and June Almeida, had used the viral strain 229E, isolated and characterized as an RNA virus a year earlier by Dorothy Hamre at the University of Chicago, from student volunteers suffering from respiratory infections [3]. Tyrrell was for a long time the head of the Common Cold Unit, Salisbury, UK, set up immediately after the Second World War in 1946 to find ways of mitigating the loss of productive work due to "colds" [4]. The realization that coronaviruses contributed substantially to human respiratory infections led Tyrrell, in 1992, to suggest that "*a programme to search for an anti-coronavirus drug might be well worth undertaking. Such a search might become more rational than in the past. Perhaps a specific binding site might be found in an essential virus protein, just as had been done for rhinoviruses*" [5]. Ironically, Tyrrell wrote with some finality in a textbook of Medical Microbiology in 1996 that "*Coronaviruses cause acute, mild upper respiratory infection (common cold)*" [6]. With far more threatening viruses emerging, most notably the human immunodeficiency virus (HIV), the causative agent of AIDS, in the last decades of the twentieth century, coronavirus research was largely relegated to the backwaters of virology. With animals, primarily bats, providing ample reservoirs of these animal viruses, the inevitable zoonotic transfer did happen. The SARS-CoV-1 outbreak in Vietnam and Southern China in 2002–2003 was largely contained due to the heroic efforts of the Italian physician Carlo Urbani (1956–2003), who recognized that he was dealing with an unusually contagious infection and persuaded authorities in Vietnam to impose extraordinary quarantine measures [7]. Sadly, Urbani contracted the infection and died shortly thereafter [8]. This encounter with a more lethal coronavirus, together with the MERS outbreak in 2007, sparked considerable research on a hitherto poorly investigated class of viruses [9–11]. It is this body of knowledge that provided the foundation for the rapid progress of research on the SARS-CoV-2 virus during the Covid-19 pandemic, a reminder that basic scientific understanding is a prerequisite for translational success. This large body of coronaviral research also included controversial attempts to engineer viruses in the laboratory, which would make them more adaptable to human hosts [12], resulting in the prolonged and inconclusive debate on the natural versus laboratory origins of the SARS-CoV-2 virus [13].

This commentary is confined to the consideration of the evolution of the surface glycoprotein of the virus during the pandemic as it provides a striking example of sequence adaptation as the virus evolves in diverse human hosts, with genetically determined immune responses that may differ, and over time periods when human populations have been immunized with different vaccines.

1.1 Visualizing the Virus Spike Glycoprotein

The grainy electron micrographs of the mid-1960s have given way to far more detailed pictures of the virus as a nearly spherical particle of about 100 nm in diameter, whose surface is dotted with the projecting spikes of the surface glycoprotein. The spike protein was the target of much of the vaccine and drug development programs in 2020 and 2021, as the SARS-CoV-2, the causative agent of the Covid-19 pandemic, spread relentlessly across the world. The ancestral sequence that emerged from Wuhan, China reveals a polypeptide of 1273 amino acids (UniProt id: P0DTC2). Three-dimensional structures were quickly determined by cryo-electron microscopy (cryoEM), aided by the previously determined structure of the protein from SARS-CoV-1 [14, 15] (Fig. 1.1).

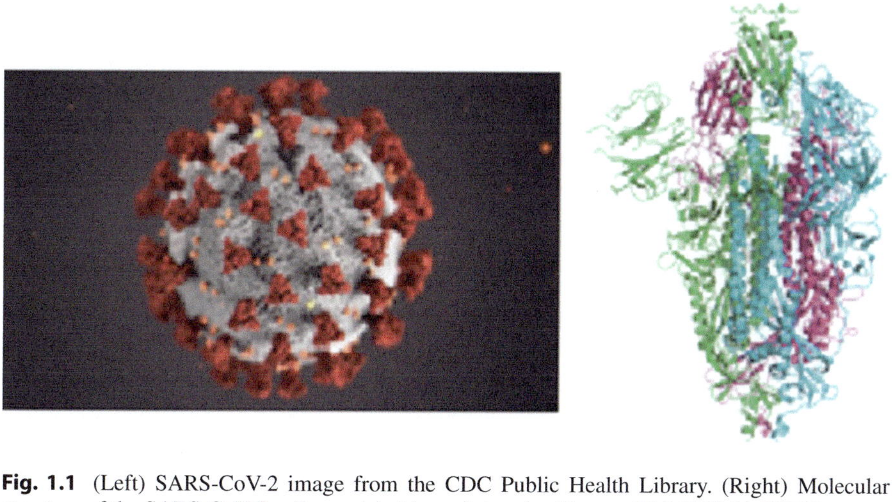

Fig. 1.1 (Left) SARS-CoV-2 image from the CDC Public Health Library. (Right) Molecular structure of the SARS-CoV-2 spike protein trimer determined by cryoEM, PDB id 6VXX

Infection is achieved by the spike protein binding to the receptor (angiotensin-converting enzyme ACE-2) on the human cell surface, facilitating membrane fusion and internalization of the viral RNA, which in turn triggers viral replication. An amazing volume of structural information on the spike protein has been accumulated in a remarkably short period of time during the pandemic years. A cursory search of the Protein Data Bank using the search term "SARS-CoV-2 spike glycoprotein" yields 20,723 experimental structures (X-ray diffraction 15709, cryo-EM 4299) [16].

Figure 1.2 shows a view of the molecular structure of the spike protein, which exists in its native state as a trimer. Examination of the monomeric polypeptide clearly reveals different structured domains. For purposes of the present discussion, it is convenient to define two domains, S1 (residues 1–695) and S2 (residues 696–1273), which separate on proteolytic cleavage at a basic site lying between residues 677 and 685. Host proteases, primarily furins, mediate this cleavage, an obligatory step before virus–host cell fusion. The protease cleavage site of SARS-CoV-2 has attracted much attention during the pandemic, fueling a contentious debate on the origins of the virus. The S1 domain consists of two distinct subdomains, the N-terminus domain (NTD, residues 1–330) and the receptor-binding domain (RBD, residues 360–580). The RBD which encodes the specificity for the

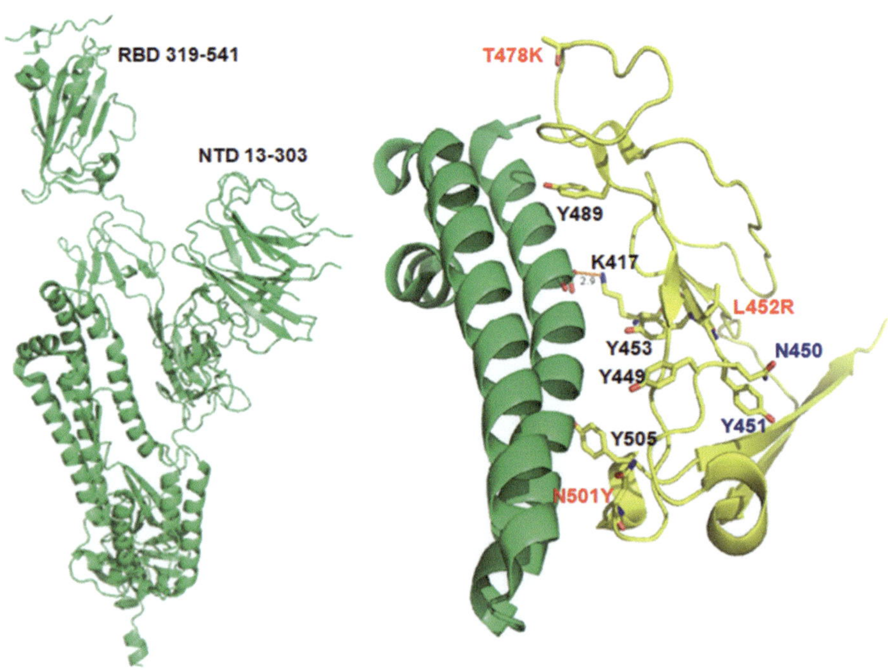

Fig. 1.2 (Left) A ribbon representation of the monomeric polypeptide chain of the spike protein showing the NTD and RBD domains, drawn using the PDB id 6VSB. (Right) A view of the dominant interactions at the interface of the RBD and the ACE-2 receptor surface drawn using the PDB id 6MOJ [17]. Highlighted residues, L452R, T478K, and N501Y were extensively discussed during the pandemic

human ACE-2 receptor has been intensively investigated, with therapeutic approaches targeting the inhibition of virus–host cell binding. Both NTD and RBD domains are involved in interchain interactions that stabilize the spike trimer and S1/S2 interdomain contacts.

1.2 Spike Protein Sequence Evolution During the Pandemic

The unprecedented volume of sequencing of viral isolates from across the world provided valuable insights into sequence evolution of the spike protein as SARS-CoV-2 adapted to human hosts and responded to pressures imposed by host immunity, both natural and vaccine-induced. The discussion that follows is based on sequences extracted from the NCBI database and is primarily based on United States data. In retrospect, two major mutational changes drove the pandemic. The first, D614G, appeared early, in the first few months of 2020, transforming the ancestral Wuhan strain into a more rapidly spreading variant. This apparently innocuous replacement of an aspartic acid (Asp, D) residue by a glycine (Gly, G) residue appears to dramatically facilitate virus–receptor interactions by favoring a more exposed conformation of the RBD domain of the virus [18]. The second mutation, P681R/P681H, appeared toward the end of 2020, fueling the spread of the alpha and delta variants which drove the pandemic in 2021. This critical mutation introduces an additional basic residue at the furin cleavage site, enhancing the efficiency of the proteolytic separation of the S1 and S2 domains, which is essential for membrane fusion [19]. Figure 1.3 summarizes the rise and fall of the viral variants during the pandemic.

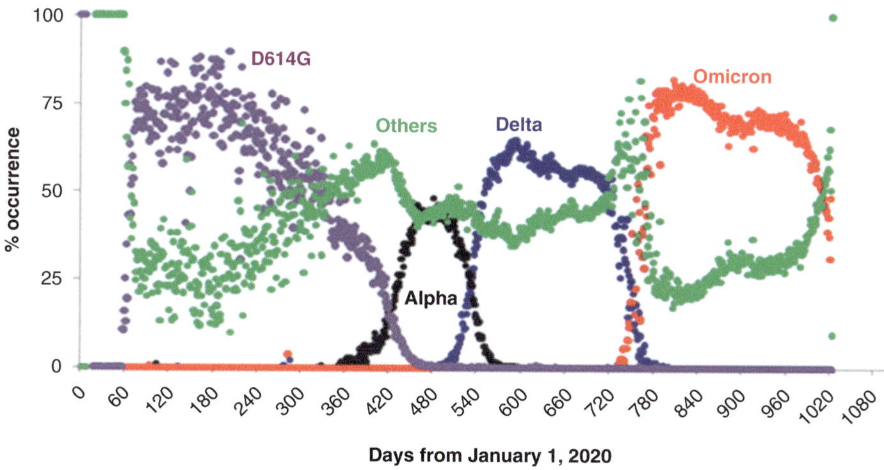

Fig. 1.3 Viral variants monitored over the period from January 1, 2020, to November 30, 2022, from the NCBI database of spike protein sequences. A total of 1,647,165 sequences were used eliminating sequences that were incomplete or partial. For the omicron sequences, the group with the 214EPE insertion is not included. The dates correspond to the sample collection dates listed in the NCBI database

Figure 1.4 summarizes the time course of appearance of the specific mutation D614G and the furin cleavage site mutation P681H/R. The mutations at residue 681 are characteristic of the alpha (681H) and delta (681R) variants. These mutations were widely discussed at different points in the period 2021–2022. While the pandemic was largely driven by the D614G mutation in much of 2020, the explosive growth of the alpha and delta variants in 2021 did most of the damage, until the emergence of the omicron variant toward the end of 2021. It is evident that the early phase of Covid-19 during the period January–March 2020 was driven by the establishment of the D614G mutant as the dominant viral strain (Figs. 1.3 and 1.4). This mutation, which undoubtedly drove adaptation to the human host, was subsequently fixed in all the dominant variants that evolved during the pandemic. The

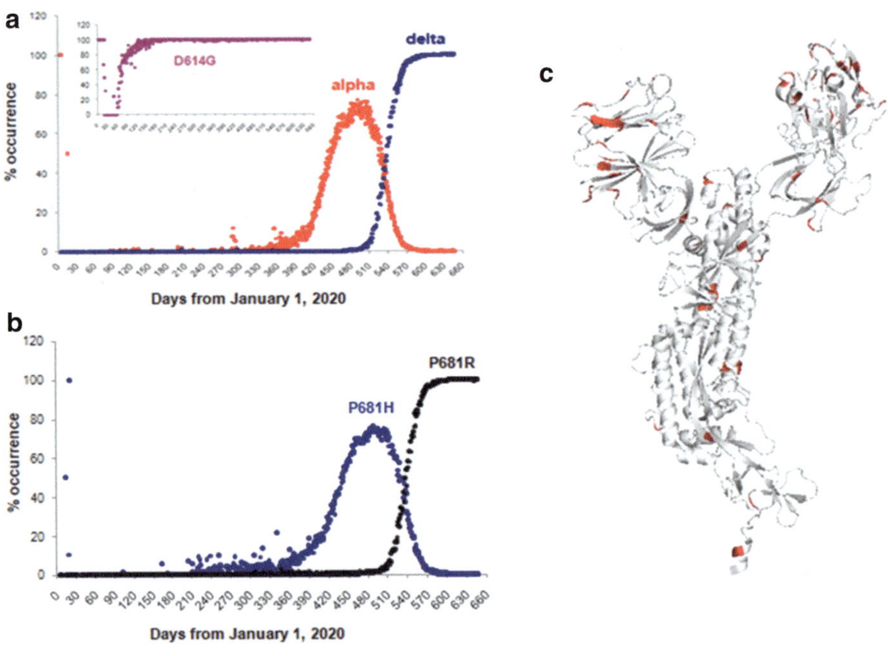

Fig. 1.4 (**a**) Incorporation of the D614G mutation in viral sequences during the pandemic. The time course of appearance of the alpha and delta variants in the period January 2020 to October 2021. Sequences containing the following sets of mutations were used for classification: alpha H69Δ, V70Δ, Y144Δ, P681H; delta E156Δ, F157Δ, R158G, P681R. (**b**) The transition between the P681H (alpha) and P681R (delta) mutations. (**c**) Sites of mutations in the major variants (see Fig. 1.5) mapped (red) on the structure of a monomeric spike protein chain (PDB 6VXX). Electron density is not observed for the following residues: 1–26, 70–79, 144–164, 173–185, 246–262, 445, 469–488, 502, 621–640, 677–688, 828–853, and 1148–1273

first signs of the alpha variant, which has the characteristic deletions at residues 69, 70, and 144, together with the P681H mutation appear in early 2021. An analysis of limited Indian data publicly available in NCBI in 2021 shows that the alpha variant was quickly overcome by the more rapidly proliferating delta strain. The delta variant is characterized by deletions at E156 and F157, together with the mutations R158G and P681R. In the US data shown in Fig. 1.3, by mid-2021, both variants coexisted in the population. Major deletions are observed in the N-terminus domain (NTD) in both variants, which are however distinct. The deletions E156Δ, F157Δ, and the R158G mutation in the delta variant point to a shedding of bulky, recognizable amino acid side chains, possibly a consequence of immune pressure-induced natural selection. The tendency of deletions in the NTD to be observed in dominant, circulating viral strains is also seen in the less lethal, but rapidly spreading strains that emerged in 2022, following the appearance of the omicron mutant. The deletions of residues 24–26 were characteristic of the BA.1/BA.2 variants that followed omicron. While the alpha and delta variants predominantly harbored 10 mutations, the omicron variants quickly displayed around 30 mutations or more. It should however be borne in mind that in the strains which spread rapidly in the human population, additional mutations accumulate in variants which belong to the same class. Viral evolution by recombination events presumably drives large discontinuous jumps in the number of mutations, with dramatically changing phenotypes.

At the time of writing (January 2024), a new variant JN.1, harboring as many as 59 spike mutations, appears to hold sway, having taken over from the variant BA2.86. Figure 1.5 summarizes the major mutations in some coronaviral variants that persisted for a significant period and drew attention during the pandemic period of 2020–2022 and highlights mutations in the three most recent variants that have been discussed in 2023. Notably, while the JN.1 variant has the P681R mutation, the XBB1.5 and BA2.86 variants carry the P681H mutation. The deletions of H69Δ/V70Δ/Y144Δ, characteristic of the alpha variant, have been re-established in the more recent strains prevalent in 2023. Variants that were discussed in early 2021 but did not propagate substantially include the epsilon, beta, and gamma (P1) variants, none of which carry mutations at the P681 residue, in the furin cleavage site.

This preliminary survey of historical spike protein sequence data during the pandemic suggests that a more detailed postmortem of mutational changes during the 4 years of SARS-CoV-2 evolution in human hosts may provide insights into the nature of the selection pressures that drove virus adaptation to human hosts. The key drivers of the pandemic appear to be mutations that promote the processes of host receptor recognition and viral–host membrane fusion, which require cleavage of the S1 and S2 domains, permitting large-scale conformational transitions in the S2 domain, an essential prerequisite for the fusion process.

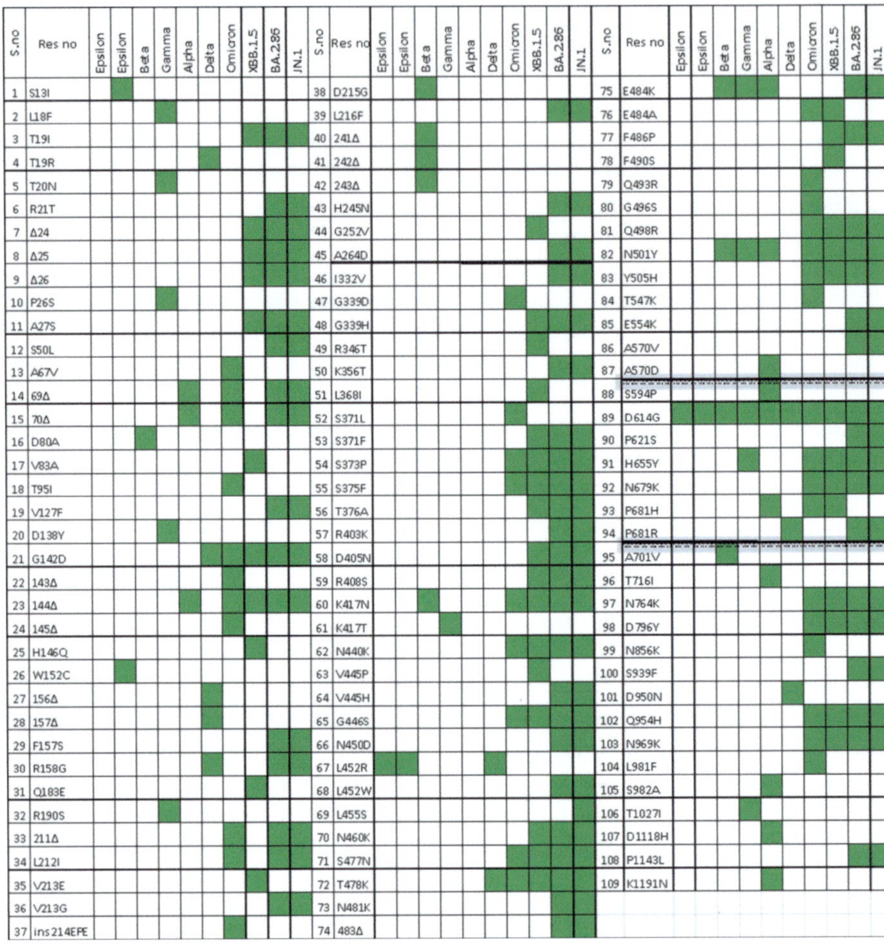

Fig. 1.5 A comparison of mutations in the spike protein in 10 variants of the SARS-CoV-2 virus. The black horizontal lines demarcate the protein structural domains. Mutations 1–45 (NTD), 46–87 (RBD), 88–94 (protease cleavage domain), and 95–109 (S2, fusion domain)

Acknowledgments M.V thanks the Jawaharlal Nehru Centre for Advanced Scientific Research for fellowship support.

Competing Interest The authors have no conflicts of interest to declare that are relevant to the content of this chapter.

References

1. News and views, virology: coronaviruses. Nature. 1968;220:650.
2. Almeida JD, Tyrrell DAJ. The morphology of three previously uncharacterised respiratory viruses that grow in organ culture. J Gen Virol. 1967;1:175–8.
3. Hamre D, Procknow JJ. A new virus isolated from the human respiratory tract. Proc Soc Exptl Biol Med. 1966;121:190–3.
4. Tyrrell DAJ, Fiedler M. Cold wars: the fight against the common cold. Oxford University Press; 2002. p. 268.
5. Tyrrell DAJ. A view from the common cold unit. Antivir Res. 1992;18:105–25.
6. Tyrrell DAJ, Myint SH. Chapter 60: Coronaviruses. In: Baron S, editor. Medical Microbiology. 4th ed. Galveston: University of Texas; 1996.
7. Reilley B, et al. SARS and Carlo Urbani. N Engl J Med. 2003;348:1951–2.
8. Cherry JD, Krogstad P. SARS: the first pandemic of the 21st century. Paediatric Res. 2004;56:1–5.
9. Li F. Structure, function, and evolution of coronavirus spike proteins. Ann Rev Virol. 2016;3:237–61.
10. Cui J, Li F, Shi Z-L. Origin and evolution of pathogenic coronaviruses. Nat Rev Microbiol. 2019;17:181–92.
11. de Wit E, van Doremalen N, Falzarano D, Munster VJ. SARS and MERS: recent insights into emerging coronaviruses. Nature Revs Microbiol. 2016;14:523–34.
12. Menachery VD, et al. A SARS-like cluster of circulating bat coronaviruses shows potential for human emergence. Nat Med. 2015;21:1508–13. https://doi.org/10.1038/nm.3985 (corrected 20 November 2015).
13. Harrison NL, Sachs JD. A call for an independent inquiry into the origin of theSARS-CoV-2 virus. Proc Natl Acad Sci U S A. 2022;119:e2202769119.
14. Wrapp D, Wang N, Corbett KS, Goldsmith JA, Hsieh C-L, Abiona O, Graham BS, McLellan JS. Cryo-EM structure of the 2019-nCoV spike in the prefusion conformation. Science. 2020;367:1260–3.
15. Walls AC, Park YJ, Tortorici MA, McGuire AT, Veesler D. Structure, function, and antigenicity of the SARS-CoV-2 spike glycoprotein. Cell. 2020;181:281–92.
16. Neuman BW, Adair BD, Yoshioka C, Quispe JD, Orca G, Kuhn P, Milligan RA, Yeager M, Buchmeier MJ. Supramolecular architecture of severe acute respiratory syndrome coronavirus revealed by electron cryomicroscopy. J Virol. 2006;80:7918–28.
17. Lan J, Ge J, Yu J, Shan S, Zhou H, Fan S, Zhang Q, Shi X, Wang Q, Zhang L, Wang X. Structure of the SARS-CoV-2 spike receptor-binding domain bound to the ACE2 receptor. Nature. 2020;581:215–20.
18. Yurkovetskiy L, et al. Structural and functional analysis of the D614G SARS-CoV-2 spike protein variant. Cell. 2020;183:739–51.
19. Saito A, et al. Enhanced fusogenicity and pathogenicity of SARS-CoV-2 Delta P681R mutation. Nature. 2022;602:300–6.

2 Laying the Roadmap for a Robust R&D Ecosystem

Renu Swarup

The spread of the disease Covid-19 across the globe, since its first outbreak in Wuhan, China, led to the World Health Organization classifying it as a pandemic on March 11, 2020, within 3 months since the first reports of the illness.

This was the first pandemic after over 100 years, caused by a new coronavirus, and as such, the world population was unaware about this pathogen. Global air travel, the mechanism of spread, a long incubation time of the virus, and asymptomatic cases also being carriers, contributed to the speed with which the virus spread.

The first case of Covid 19 in India was confirmed on January 30, 2020. This global pandemic caught the world unaware and the global scientific and medical communities were at the frontline of mitigating the damage from this disease, by understanding more about the pathogen, tracking its evolution, and developing and implementing treatments and diagnostics. This was a global call for action that the scientific community responded to. There were several concerted and individual global efforts to develop drugs, vaccines, and diagnostics to address the challenges posed by the disease, and governments and organizations across the globe collectively committed themselves to this task.

In India, the scientific community also came together to collaborate, in national as well as international efforts, to develop solutions to the pandemic. The S&T strategy in India against Covid-19 relied on four pillars—understanding the pathogen and the disease, identifying the affected patients through efficient and accurate diagnostics, developing cures through therapeutic agents, and developing vaccines for the prevention of the disease.

It is important to look back to see how we began—it was late 2019 December when we were all suddenly informed about a novel Sars Cov 2 strain which was causing widespread infection in China. Having started in Wuhan, early reports indicated that this is a

R. Swarup (✉)
Department of Biotechnology, Former Secretary to Government of India, Ministry of Science and Technology, New Delhi, Delhi, India
e-mail: renuswarup01@gmail.com

© The Editor(s) (if applicable) and The Author(s), under exclusive license to Springer Nature Singapore Pte Ltd. 2025
R. Swarup et al. (eds.), *At War with the Single Strand*,
https://doi.org/10.1007/978-981-96-8122-8_2

Coronavirus strain and is highly infectious; however, the exact origin of this was not clear. Since we have had an earlier experience of SARS, there were immediate concerns within scientific research groups and public health specialists on managing the infections through scientific tools and a well-developed public health system. As we moved into January, we saw the rate of infection increasing. The number of cases in China and in other countries across the world were increasing. We began our discussions on what would be the possible mechanisms for managing this as a Public Health emergency and also what are the options that science will have to put forward to fight this virus. Being a SARS virus, it was a single-strand RNA virus, and there were earlier examples of developing therapeutics, diagnostics, and vaccines to control such outbreaks, however, to control a small number of infections. In India, we were fortunate till that time this virus had not entered our country in a large way, but by that time, the health officials and the scientific research community had started working on a very robust strategy to tackle this. We had geared up our scientific research groups, our industry, the entire ecosystem, and all stakeholders to come together and chalk out a road map on what the scientific strategy should be. The first R&D strategy meeting was organized by the Department of Biotechnology in mid-January.

Our R&D strategy was prepared focusing on four major aspects (Fig. 2.1).

(i) Understanding the virus—Genome sequencing.
(ii) Testing—Diagnostics, especially molecular diagnostics.
(iii) Treatment—Novel therapeutics, natural products, and phytopharma.
(iv) Prevention—Development of novel vaccines.

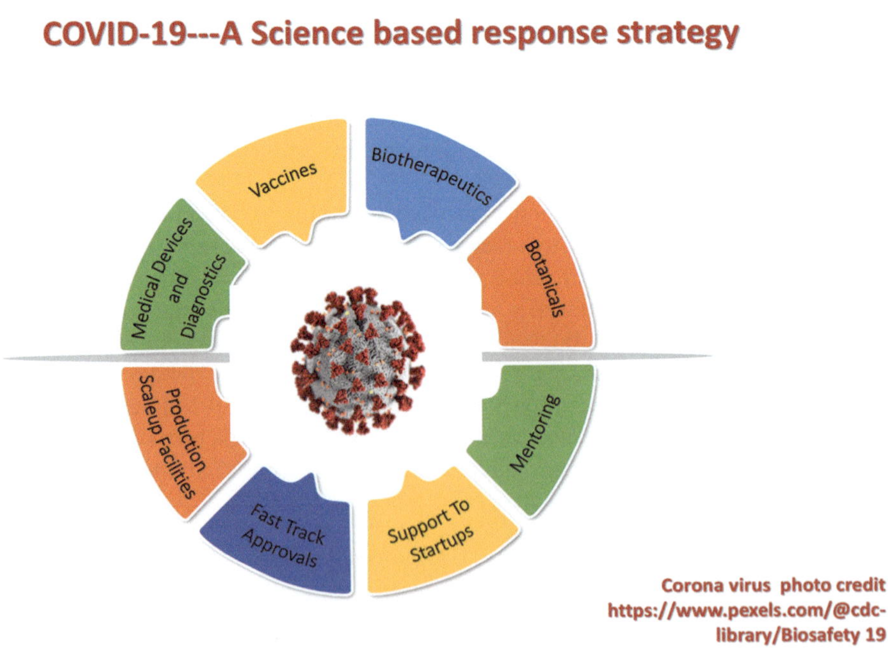

Fig. 2.1 The COVID response strategy

The first R&D blueprint discussion by WHO had the participation of Indian scientists. We presented our action plan to WHO in February 2020, and by that time we had already begun working on the different aspects of diagnostics, therapeutics, and vaccine development.

Globally, we saw that as soon as the sequence was released, the international collaborators began discussing on how the viral sequence information which was available should be used for developing vaccines, what are the diagnostics that we need to develop, and what are the therapeutic options that we have. It was clear that we will have to bring together all the existing strength and competence to take on this urgent task of developing medical countermeasures in the shortest period of time. The virus was moving with great speed, and our scientific development had to beat this.

2.1 The Governance and Decision-Making Framework

The whole of government approach was adopted, and there was a very high-level involvement by the country's leadership to take speedy actions and decisions. The PMO constituted special empowered groups, and one of them was on Medical Emergency and Management Plan chaired by Dr. Vinod Paul, Member Niti Aayog, which discussed the scientific aspects of the management of COVID. A report on the scientific action plan for managing this disease was prepared and presented for necessary action by concerned agencies.

A special task force on COVID-19 vaccines and S&T solutions was set up, chaired by Member Niti Aayog and Principal Scientific Adviser with the Secretaries of all Scientific Ministries/Departments, which discussed the components of the ecosystem which need effective collaboration and focused attention. All departments and ministries came forward and collectively took on the task of supporting research in different areas. DBT, DST, CSIR, DRDO, AYUSH, and ICMR all handled specific areas. NEGVAC, which is the National Empowered group on Vaccine administration, and NTAGI, which is the National Technical Advisory Group for Immunization, closely monitored and guided the development of medical countermeasures and their effective deployment across the country.

A joint group comprising Atal Innovation Mission, BIRAC, and other concerned departments supported young start-ups to come up with innovative, affordable solutions for COVID. More than 200 different solutions have reached the market from masks N95, to PPEs, to ventilators, to oxygen supplies, to thermal scanners, sanitizers, etc.

2.2 The Virus—Understanding the Structure and Mechanism and Tracking the Evolving Mutations

The search of the causative agent soon led to the identification of novel coronavirus using metagenomic next-generation sequencing technology. Due to the genetic association of this virus with coronavirus responsible for severe acute respiratory syndrome (SARS) and Middle-East respiratory syndrome (MERS) outbreaks, it was named as severe acute respiratory syndrome coronavirus 2 (SARS-CoV-2)

[1]. World Health Organization (WHO) termed this respiratory illness as COVID-19. Being an RNA virus, it was evident that with increasing number of passages, there would be mutations, and it is therefore important to sequence the viral genome to identify new variants and clinically correlate these to assess the level of infection and pathogenicity. Initially, individual efforts were launched in the country covering different geographic locations to sequence the Sars-Cov2 genome. DBT, CSIR, and some state- level organizations began sequencing the viral genome, and we had about 2000 genomes sequenced, which gave us an insight into the level of mutation. Globally, the viral sequences were being deposited in the global database GISAD, and Indian scientists also began depositing the sequences. In November/December 2020, there were reports of a highly infectious strain of clinical concern reported initially from UK and then very soon from many other countries. WHO notified this as the Alpha strain—a variant of concern. It was important to track the presence of this in our country to be prepared to tackle the increasing levels of infection and the clinical management of increasing hospitalization cases. An Indian SARs-Cov 2 Genome Sequencing Network—INSACOG was set up on priority. Steered by the Department of Biotechnology and Ministry of Health, with the participation and deep involvement of NCDC, ICMR, CSIR, and state laboratories, within days of the first Alpha strain being reported in India, a network of 10 laboratories began aggressive screening of samples and analzsing the genomic data to study the transmission of this highly virulent strain. This was clinically correlated and provided the scientific data for rapid response in the emerging hot spots. INSACOG expanded to then include more labs, and Indian data was also being provided to GISAD. The deadly Delta strain and subsequently Omicron and its subvariants were detected by the INSACOG network. Not only was this data useful for managing the disease but also helped researchers to develop new diagnostics and vaccines which could respond to these new strains. A hospital network and sewage surveillance network was connected so that the genomic data could be correlated with epidemiological surveillance data on an urgent basis and provide an early warning system.

INSACOG is today a network of more than 65 laboratories that provide a robust genomic surveillance network which is important not only for COVID but other priority pathogens also. COVID resulted in the building of sequencing capacities across all national and private laboratories, both in terms of infrastructure and human resource. A highly interactive portal and integrated data system was built to allow smooth movement of data for a quick decision-making process. This capacity can now be leveraged to build surveillance networks for early detection and warning of pathogens.

2.3 Call for Action for Developing Medical Countermeasures

The DBT along with the public sector BIRAC launched the DBT—BIRAC Covid Research Consortium, as early as March 15, 2020, and supported the time-bound innovation research for urgent product development of affordable, indigenous diagnostics, therapeutics, and vaccines. The need to support a vibrant and robust ecosystem for translational research product development was considered and the existing gaps identified. It was evident that there was an urgent need to bring all players from academia, industry, and start-ups together,

and quick decisions were required to provide the necessary support. The DBT/ BIRAC Consortium was guided by a high-level expert group which provided quick, transparent, and technically strong decisions. High-risk research and innovation were supported for high returns. Collaborations were forged, existing strengths were leveraged, and existing gaps in the ecosystem were identified and strengthened. There was only one key objective and aim—the need for indigenously developed, affordable medical countermeasures which can meet the requirement of over a billion in the shortest timeframe. The focus was on building and strengthening a strong robust ecosystem required for the development of the needed medical countermeasures. This included the development of new platform technologies, animal models, shared infrastructure, development of assays, testing labs, biorepositories, and clinical trial network sites.

2.4 Testing and Tracking—Building a Diagnostic Supply Chain

The diagnostic development is a success story which impacted every citizen in the country by making indigenous diagnostic kits available in large numbers at affordable costs. In March, when we started testing in a big way, we were totally dependent on imported kits. We put out a challenge before our research groups, start-ups, and industries that we need indigenous kits for two reasons—first, of course, we wanted to be self-reliant within the country to respond to the call of Atma Nirbhar Bharat, and second, it was also important to develop our indigenous capacities, since there were supply-chain disruptions, and the global demand for diagnostic kits was huge. The challenge of developing indigenous capacities was taken up by our start-ups and researchers, and within a period of less than 60 days, we had diagnostics developed to cater not only for the nation but also the global market. For RTPCR, antibody test, and antigen tests, more than 100 indigenous kits were developed by our innovators (Fig. 2.2).

The DBT set up the command consortia at the AMTZ, Vishakapatnam. This is a plug-and-play facility which allows many of the diagnostics group to go and immediately start their production. All facilitation was provided, and this was of course in addition to the extraordinary work that was done for the development of indigenous PPEs, masks, and ventilators at the AMTZ. The facility could produce over 1 lakh tests per day and also required reagents (Fig. 2.3). We also built the first **I-Lab—**the mobile infectious disease laboratory—which is not just for sample collection but also for diagnostic testing. The first mobile lab was positioned at the DBT-Translational Health Science Institute at Faridabad and catered to the remote districts of Haryana. Three more such facilities were set up and deployed in far-flung areas of Northeast and other parts of the country.

Fig. 2.2 COVID 19 Made in India diagnostic kits

Fig. 2.3 Manufacture of COVID 19 diagnostic kits

What was also important was that to develop the diagnostic kits indigenously, we also required the components, reagents, etc. For this, we quickly brought together **N-BRIC**—the National Biotech Resource Indigenization Consortium which was led by DBT. Dr. Kiran Mazumdar and her team from ABLE came forward, and C-CAMP, our incubator at Bangalore, took on the lead. More than 30 manufacturers who were producing reagents, etc. for other applications were brought together under NBRIC and encouraged for producing different reagents required for our diagnostic kits. There was a demand aggregation, and the consortia ensured a confirmed supply to the diagnostic manufacturers. We successfully developed indigenous kits including the components and reagents. This was a very clear evidence of how global collaborations, national collaborations, and intersectoral collaborations can move forward for a common cause, for a common goal, and successful delivery on schedule.

2.5 Vaccines—Saving Lives, Fighting the Pandemic

Then of course comes our wonderful story on vaccine development; never before had we ever thought that in a period of less than 12 months, we would have vaccines developed and approved for human administration. The capacity and competence of India as a vaccine manufacturer is globally recognized. 60% of the world's immunization vaccines are produced within the country. But, here it was a challenge of COVID, and our vaccine developers all came together, quickly started their work, and it was a matter of great pride and satisfaction when two of our vaccines got EUA in less than a year of development. India was one of the few countries which supported a robust pipeline of vaccine candidates. The DBT–BIRAC COVID consortia supported more than 30 vaccine candidates involving academic institutes, research laboratories, small industries, and vaccine R&D companies. High-risk funding was provided for new technology platforms. We facilitated the translational research by creating a very strong ecosystem.

It was this robust ecosystem that made India the only country to proudly have a basket of vaccine candidates under development, from different technology platforms. The first two vaccines (antibodies), Covishield (S subunit antigen) and Covaxin (inactivated whole virus antigen), received EUA in January 2021. Thereafter, we had the first DNA vaccine Zycov-D, first thermostable mRNA vaccine, and the first nasal vaccine; the subunit vaccine Corbevax has also recently received the WHO authorization.

The National Biopharma Mission, a Government of India partnership with World Bank launched by DBT/BIRAC in 2017, has focused on *Innovate in India* and has built a unique innovation ecosystem, bringing together industry, academia, and start-ups. A state-of-the-art network of shared infrastructure has been created across the product development value chain for all Biopharma products; this was the foundation for our success in the pandemic.

The other major mission approved by the government and launched by DBT / BIRAC in 2018, the IndCEPI Mission, which was the India component in partnership with CEPI, the organization supporting vaccine development for epidemic preparedness, had already established a very sound foundation to build the Biopharma Indigenous Innovation ecosystem, and this accelerated our preparedness for vaccine research and development. We already had key components available in terms of shared infrastructure facilities and trained human resource also as research resources, which very quickly could be taken up for the work required to facilitate COVID vaccine development.

To strengthen the Indian COVID vaccine development program, **Mission COVID Suraksha** was launched under the AtmaNirbhar Bharat 3.0 Package. This was the country's first national mission focusing on an end-to-end vaccine development from clinical development and clinical trial to manufacture and regulatory approval. The aim was to accelerate the development of at least five to six vaccines, taking them from preclinical to phase 1 to phase 3 clinical trials to market licensure and also building and supporting the ecosystem for indigenous research capacities within our country.

2.6 Key Components of the Mission Focused On

2.6.1 Development of Resources for Vaccine Development

Laboratories to conduct safety and immunogenicity animal studies: Animal models and challenge models are the essential components of preclinical studies for vaccine development. GLP- certified labs with suitable small animals (mice, rabbits, guinea pigs, and hamsters) and large animals (NHPs) were set up/strengthened in national laboratories for safety, repeat toxicity, and immunogenicity preclinical testing.

(i) Laboratories equipped to conduct animal challenge studies were identified as a gap. BSL3 facilities with access to suitable small and/or large animals in compliance with GLP and other regulatory requirements were supported in national laboratories to assess the protective ability of COVID-19 vaccine candidates prior to Phase 3 efficacy studies for COVID-19, as required by the regulators. Five of the DBT national laboratories took up the challenge of developing disease-specific animal challenge models. The first Ferrat facility was also set up at THSTI, Faridabad.

(i) Centralized Immunoassay Laboratories: GCLP-compliant labs with validated high-throughput immunoassays specific for COVID-19 with sufficient capacity to meet the demand are an essential requirement to support clinical development. Considering that a number of candidates would move into trial at the same time, it was felt necessary to support these laboratories. Two such laboratories were set up at THSTI and IRSHA. The International Coalition for Epidemic Prepardeness—CEPI—set up an international network of immunoassay labs for the international global trials of vaccine candidates. The THSTI Immunoassay Lab was one of the five labs identified by CEPI as a part of their network (Fig. 2.4).

2.6.2 Clinical Trial Site Network

The availability and access to GCP-compliant clinical sites, community-based, or hospital-based sites with access to target population for Phase 1, 2, or 3 trials are critical for the clinical development of COVID-19 vaccines. Under the National

Fig. 2.4 Bioassay Laboratory at THSTI. (Recognized by CEPI as one of the seven global networks of laboratories for harmonized assays for vaccine development)

Biopharma Mission, after a very rigorous selection, a number of sites were selected for conducting Phase 3 clinical trials. They were very quickly prepared for conducting field trials for COVID vaccine. All these sites had access to the existing cohort of around 1,00,000 plus healthy subjects (Phase 1 clinical trials). To prepare them as the most appropriate sites for trials, parallel efforts were made for conducting sero-epidemiology studies, preparing to conduct trials as per Good Clinical Practices Guidelines (GCP), with access to a Central Lab for Immunogenicity and Central Biorepository. Within a year, as the vaccine development progressed, a total of over 50 sites were identified and strengthened. This also included hospital network sites and were spread across the country.

2.6.3 Manufacturing Capacities for Manufacture, Scale-Up and Fill and Finish Facilities

1. *Process development and clinical trial lot manufacturing*
   - Capabilities for process development, cell line development, and manufacturing of GMP batches for toxicology studies in animals and clinical trials were also required. While few manufacturers have capabilities for this in-house, many outsource these to CMOs. With multiple vaccine candidates in pipeline, these facilities are critical to ensure timely entry in the clinic. Additionally, with new vaccine platforms being leveraged to develop COVID-19 vaccines, the need to establish these facilities for process development and clinical lot manufacturing was identified as a critical gap. The facilities set up under the National Biopharma Mission were leveraged and further strengthened.
2. *Manufacturing facilities for bulk scale-Up*

   We developed a very large portfolio of vaccines on different platform technologies (Fig. 2.5). Investment was also made to strengthen existing capacities for enhanced production to meet the requirements of bulk manufacture, fill and finish, and other logistics. Covaxin production capacity was increased by strengthening and setting up viral vaccine production capabilities at three locations—Bharat Biotech, Bengaluru, Indian Immunologicals, Hyderabad, and Haester, Gandhinagar.

We also created COVID-specific repositories in five locations which had clinical samples deposited, which were essentially required for all the research work that was done by the researchers from Academia and Industry—for the validation of the kits, for assay development, for vaccine candidate development, etc. The

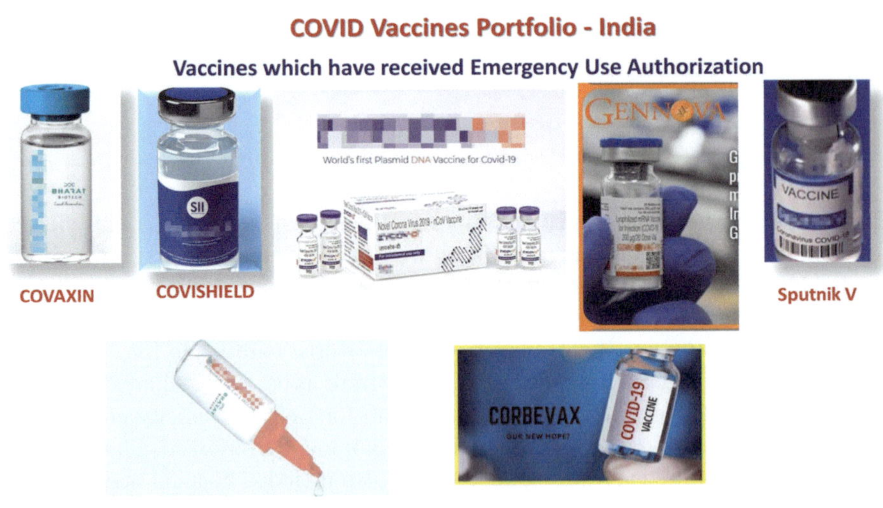

Fig. 2.5 COVID vaccine portfolio

government also issued specimen-sharing guidelines which allowed specimens to move and be shared for research.

A Rapid Response Regulatory framework was developed which helped in quickly responding to the regulatory requirements. The pathway ensured that there is complete transparency and efficiency without compromising on the quality of the data and product.

One mandatory requirement for vaccine batches to be released in the market for human immunization is their clearance by the vaccine testing laboratory as per the CDSCO guidelines. There was only one vaccine testing laboratory in the country at Kasauli. This laboratory has very specialized infrastructure and human resource. As the vaccine development progressed in the country, it was evident that we will have millions of doses produced each month; the need for more vaccine testing laboratories was recognized. Very quickly, in about less than 3 months, two more vaccine testing labs were set up at the DBT Institutes—NCCS, Pune and NIAB, Hyderabad. This has now enhanced our vaccine production and delivery.

Today, we are at a position where in the area of vaccine development, we are confidently catering to the requirements of vaccines for India and vaccines for the world. Under the Vaccine Maitri Initiative, India supplied a large number of doses to over 100 countries—a soft power diplomacy. We were also one of the largest supplier to COVAX.

Under the PACT—Partnership for Accelerated Clinical Trial—we also worked toward building clinical trial capacities within our neighboring countries. Health care- and clinical trial-related workers were trained in more than 10 countries.

2.7 Strong Academia–Industry Interface and International Collaborations

The ecosystem built over the last few years and strengthened at the time of COVID also had a strong component of collaborations—both national and global and also industry–academia. These partnerships which had been built over many earlier programs were further strengthened during COVID. The strong industry–academia partnership was instrumental in our first indigenous vaccine, Covaxin, being developed. Other vaccine candidates developed also had strong components of public–private partnerships. The Indo US VAP program, in over 30 years of collaboration, had provided very strong support for many earlier vaccine initiatives. For the COVID mission, once again, Indo US VAP provided mentorship, valuable advice, and access to biological resources and infrastructure.

2.8 Conclusions

Clearly, this has been a year of the Pandemic but also clearly a year of Science. We are very proud of how our science has responded, and we would like to congratulate the scientists. Moving forward, this is now going to be not just an ecosystem we have built for COVID but an ecosystem much beyond that. So that is really what the year 2020 has been, and we are confident that moving forward, it is going to be now stronger, a much stronger science-based ecosystem responding to all other challenges. Our preparedness for future is there, and we need to ensure that this further strengthens.

For a successful scientific ecosystem to deliver, there are three important pillars—capacity, cutting-edge technologies, and collaboration. The country has seen a special focus on all these three during the pandemic. It is, however, important that this robust ecosystem, which has delivered so effectively in this crisis, continues to do so for future challenges.

The country today is proud of its Biotech start-up ecosystem, and the recent trends of private investments in high-risk technologies are encouraging. We now see many start-ups focusing on cutting-edge technologies such as cell and regenerative medicine, organ culture, synthetic biology, gene editing, artificial intelligence, and the whole area of "omics". The increased investments in high-risk research and early- and late-stage financing of technology start-ups are done because of the high level of confidence in the quality and timely delivery of technologies and products by the Biotech sector. It is also important to note that there is a high level of risk mitigation, with the government becoming an equal partner in most high-risk research ventures.

There are now many other healthcare interventions which need urgent science-based solutions. Antimicrobial resistance (AMR), with a focus on One Health issues, is a silent pandemic which is slowly but stealthily spreading across the world. Recent data released in the Lancet report (January 2022) and figures by WHO on the mortality levels due to AMR are alarming—an estimated number of 4.95 m deaths in 2019 in 204 countries are associated with bacterial AMR and 1.29 m are as a result of the resistance factor. It is imperative that the network we have built for COVID and our robust knowledge translation ecosystem is expanded, and similar solutions are developed to address this silent Pandemic which is impacting both humans and animals, especially focusing on zoonotic diseases.

The diagnostic success story needs to be replicated for other important diseases. What we now need is to scale-up the COVID model, utilizing effectively the well-established clinical validation network for other diagnostics and devices.

The learning experience during COVID gives us an opportunity to make the best possible use of this ecosystem, knowledge, and capacity for other innovative novel products of national and global priority. Our scientists and researchers have very confidently built this ecosystem block by block. Our innovations are affordable and accessible and ready to respond to the needs of the nation and be replicated across borders.

The country has responded to the call of **Atmanirbhar Bharat**—a self-reliant India with innovations for India and the world. When COVID struck, developed countries projected India as a soft target, awaiting a major disaster. Of course, there were many deaths but less than that in developed countries. But India quickly addressed deficiencies in health care, at least to handle the COVID onslaught. We are now Future-Ready—Hopefully, a tiny virus cannot now catch us off-guard!

There is another area, which India is capable of assuming leadership. While COVID vaccines and diagnostics became prominent, there is no specific drug as such against this virus. MNCs have announced a few potential drugs. But, antiviral drugs for COVID are a challenge, and the area needs attention. The strength of India has been diverted to generics (drugs out of patent), and it is important to strengthen the area of drug discovery and development. While the private sector in general has a dominant role, it is necessary for the government and foundations to step in. Decades ago, India was projected as a potential leader in drug discovery and development but got diverted to generics for economic considerations. It is necessary to fulfill this promise not only for COVID but also to handle major afflictions, both infectious and chronic diseases, including diabetes and cancer.

Competing Interest The authors have no conflicts of interest to declare that are relevant to the content of this chapter.

Reference

1. Coronaviridae Study Group of the International Committee on Taxonomy of Viruses. The species severe acute respiratory syndrome-related coronavirus: classifying 2019-nCoV and naming it SARS-CoV-2. Nat Microbiol. 2020;5(4):536–44. https://doi.org/10.1038/s41564-020-0695-z. PMC 7095448. PMID 32123347.

Understanding the Virus: The Evolving Mutations

3

Saumitra Das, Sudhanshu Vrati, Arindam Maitra, Nidhan K. Biswas, Anindita Banerjee, Anup Mazumder, and Sreelekha Dutta

In the recent past, the world witnessed the harrowing pandemic of SARS-CoV-2. Being the second most populous country, the effective strategy taken by the Indian government to fight back is truly praiseworthy. As we emerge out victorious, it is time to look back and understand how we won the battle, which was fought bitterly and with substantial casualties. It is particularly important to understand how 1.4 billion people of India, with health and other resources expected to stretch thin, could counter the challenge successfully. The underlying strategic policies, the scientific initiatives, and the huge response which fueled this success should be assimilated thoroughly as these will help us in future outbreaks.

During the end of 2019 (December), cases with pneumonia-like symptoms of unknown origin started appearing in Wuhan, Hubei Province, China. The first case could have originated from the live animal-selling Huanan Seafood wholesale market. Subsequently, on January 7, 2020, the Chinese health department designated the cause of this outbreak to a novel coronavirus, and this was eventually named as Severe Acute Respiratory Syndrome Coronavirus-2 (SARS-CoV-2). There were some similarities with the genetic constitution of the Severe Acute Respiratory Syndrome (SARS) virus responsible for the 2002–2003 outbreak. On January 11, China reported its first death from the virus. The virus quickly spread within China, and cases started appearing in other countries. The World Health Organization

S. Das (✉)
National Institute of Biomedical Genomics, Kalyani, West Bengal, India

Department of Microbiology and Cell Biology, Indian Institute of Science, Bengaluru, Karnataka, India
e-mail: sdas@iisc.ac.in

S. Vrati
Regional Centre for Biotechnology, Faridabad, Haryana (NCR Delhi), India

A. Maitra · N. K. Biswas · A. Banerjee · A. Mazumder · S. Dutta
National Institute of Biomedical Genomics, Kalyani, West Bengal, India

© The Editor(s) (if applicable) and The Author(s), under exclusive license to Springer Nature Singapore Pte Ltd. 2025
R. Swarup et al. (eds.), *At War with the Single Strand*,
https://doi.org/10.1007/978-981-96-8122-8_3

(WHO) declared it to be a Public Health Emergency of International Concern (PHEIC) by January 30. On February 11, 2020, the WHO named the disease COVID-19, and by March 11, 2020, the WHO declared COVID-19 a global pandemic as the virus was rapidly spreading worldwide, affecting numerous countries, and leading to widespread illness and death [1, 2].

SARS-CoV-2 is a single-stranded, enveloped RNA virus that belongs to the family Coronaviridae. The virion particle has a spherical or pleomorphic (irregular) shape, often appearing somewhat spherical or oval when observed under an electron microscope (Fig. 3.1). The diameter of the virion particle is estimated to be around 60 to 140 nm. The distinctive feature of its appearance is the presence of spike-like projections protruding from its lipid envelope, which gives it a "crown-like" or "corona" appearance. These spikes are made up of viral spike (S) proteins involved in the virus attachment to the angiotensin-converting enzyme 2 (ACE2) receptor in human cells. This attachment is a critical step in the virus's ability to enter and infect host cells. The envelope (E) protein is a small transmembrane protein that contributes to the structure of the viral envelope and is involved in the virus assembly and release. The membrane (M) protein is another transmembrane protein that plays a critical role in maintaining the virus's structural integrity. It helps shape the envelope and interacts with the nucleocapsid (N) protein to package the viral RNA [3].

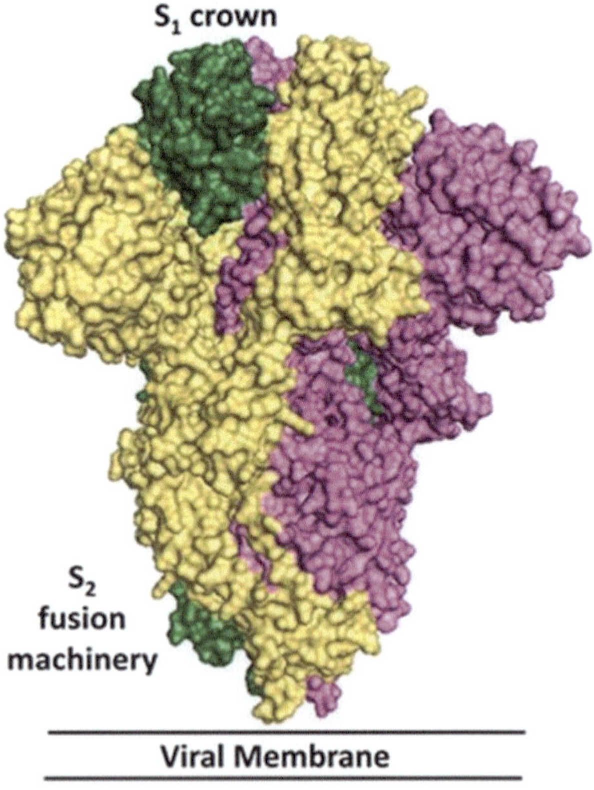

Fig. 3.1 Structure of SARS-CoV-2 spike protein. (Adopted from Maitra et al. [7])

The single-stranded RNA genome of SARS-CoV-2 contains all the genetic information necessary for the virus to replicate and produce its proteins. The genome is about 30,000 nucleotides long and encodes various structural and nonstructural proteins. The genome has untranslated regions (UTRs) at the 5′- and 3′-ends which have a key role in viral RNA replication and translation. The genome contains several open-reading frames (ORFs) that are translated into several structural as well as nonstructural proteins.

The S protein of SARS-CoV-2 is a crucial viral protein that plays a central role in the virus's ability to enter and infect host cells. It is a trimeric glycoprotein; each subunit contains two key functional domains. The S1 domain has a receptor-binding domain (RBD) that interacts with the ACE2 receptor on the surface of human cells. This binding facilitates the virus's attachment and entry into the host respiratory tract cells. After binding to the ACE2 receptor, a process triggered by various cellular and viral factors, the S2 domain of the spike protein undergoes conformational changes. These changes facilitate the fusion of the viral envelope with the cell membrane of the host, allowing the viral RNA to enter the host cell cytoplasm and initiate infection.

SARS-CoV-2, like all RNA viruses, undergoes mutations over time. Some mutations can lead to changes in the virus's behavior, including increased transmissibility, altered disease severity, or potential resistance to immunity from prior infections or vaccines. These variants of concern receive special attention due to their potential impact on public health. To monitor and understand the mutation frequency and its implications, continuous genomic surveillance efforts were crucial. Scientists and public health organizations around the world were actively sequencing viral genomes to track the emergence of different variants, which helps formulating public health strategies and vaccine development [3, 4].

Neutralizing antibodies produced against the spike protein can inhibit the binding of the virus to the ACE2 receptor, thereby blocking the viral entry. The spike protein is also prone to mutations, which can lead to the advent of new SARS-CoV-2 variants. Some variants have mutations in the RBD that might affect its interaction with the ACE2 receptor. Spike protein mutations can have various implications for the virus's pathogenesis and its interactions with the host and the immune system. While not all mutations lead to significant changes in the virus behavior, some mutations in the spike protein can impact viral transmissibility, disease severity, immune response, and even the effectiveness of diagnostics, treatments, and vaccines. Thus, surveillance and monitoring of spike mutations are crucial to identify the variants of concern and to adapt public health strategies, treatment approaches, and vaccine development efforts accordingly. Scientists and health authorities across the globe have been closely tracking these mutations to stay ahead of potential challenges posed by evolving viral variants (Fig. 3.1).

The receptor-binding domain (RBD) of the structural protein spike has a significant function in the virus's ability to bind to the ACE2 receptor and enter human cells. The RBD consists of about 200 amino acids. The specific amino acid sequence and the three-dimensional structure of the RBD are important for its interaction with the ACE2 receptor. Mutations in this region can alter the shape and charge distribution of the RBD, potentially affecting its affinity for ACE2, which in turn could affect the transmissibility of the virus and its immune evasion. There have been several mutations in the RBD of SARS-CoV-2. A few of the notable ones are listed here. The N501Y mutation marks the substitution of asparagine (N) with tyrosine (Y) at position 501. This mutation is associated with the increased binding affinity of the virus to the ACE2 receptor and is found in multiple VoCs, including B.1.1.7 (UK variant) and B.1.351 (South African variant). The E484K mutation involves the substitution of glutamic acid (E) with lysine (K) at position 484. This mutation has been associated with reduced neutralization by some antibodies, potentially affecting the efficacy of immune responses. It has been found in variants like B.1.351 (South African variant) and P.1 (Brazilian variant). The K417N/T mutations involve the substitution of lysine (K) with asparagine (N) or threonine (T) at position 417. These mutations have been observed in B.1.351 (South African variant) impacting the interaction with ACE2 and antibody recognition. The L452R mutation involves the substitution of leucine (L) with arginine (R) at position 452. It has been identified in B.1.427/B.1.429 (California variants) affecting the binding of the RBD to ACE2 [5].

3.1 The Outbreak and the First Introduction of Virus in India, Early Genome Sequencing

Countries with rich population densities have witnessed the detrimental outcomes of COVID-19 pandemic. India, which has the second highest population density globally, exhibited a slow progression of SARS-CoV-2 infection during early 2020. The first cases were reported from Kerala on January 30, 2020, by a student returning from Wuhan, China. Eventually, during the second half of 2020 and thereafter the rate of infection increased in India. By October 31, 2022, the severity of infection (44,654,638 cases and 529,077 deaths) was surpassed only by that of the United States of America (97,505,662 cases and 1,070,622 deaths) [6].

During 2020–2022, India experienced four SARS-CoV-2 pandemic waves. The first wave caused by the B.1 ancestral lineages took place between May 2020 and February 2021, while the infection surge of nearly 98,000 cases was on 1 September 16, 2020. It was followed by a comparatively more detrimental second pandemic wave, fueled by the Delta variant of concern, which had emerged in India (>23.9 M cases and > 320,000 deaths). This Delta-driven wave was seen to reach the peak on May 6, 2021 (414,188 cases). The third pandemic wave upsurged on January 14, 2022 (533,035 cases), and India witnessed a marked increase in the number of cases (~8 M cases and ~ 33,000 deaths) on account of the spread of SARS-CoV-2 variants of concern like Delta and Omicron BA.1 and BA.2 strains. This wave was ruled by the BA.2 Omicron strain and was comparatively less severe due to the effective vaccination strategies. The fourth wave (June–September, 2022) was the least severe one (~1.27 M cases and ~ 3000 deaths), predominated by the BA.2 and BA.5 Omicron variants. The VoCs Beta and Gamma, and the VoI, Mu, were sporadic in

India. The Alpha strain was detected on December 29, 2020, but was not seen to prevail much in India, while the Beta strain was seen first on March 26, 2021. Delta was reported to be the most prevalent one in India and was first reported in the state of Maharashtra on March 31, 2021. The first case of Omicron BA.1 was detected on November 28, 2021.

An early study (March 2020) on sequencing of the SARS-CoV-2 genomes was conducted with naso-oropharyngeal swabs collected from the SARS-CoV-2-infected patients across West Bengal, eastern India (led by the NIBMG research group). Nine SARS-CoV-2 samples were sequenced, and the genomic characteristics of the virus with respect to the various other sequences reported from India were compared and analyzed. Unique mutations were detected in the three protein-coding viral genes, namely, RNA-dependent RNA polymerase, spike, and nucleocapsid in the circulating D614G variants. It appeared that the nucleocapsid gene mutations is equally important since it leads to the alterations in the protein structure, impacting the pathogenicity of the virus (Fig. 3.2) [7].

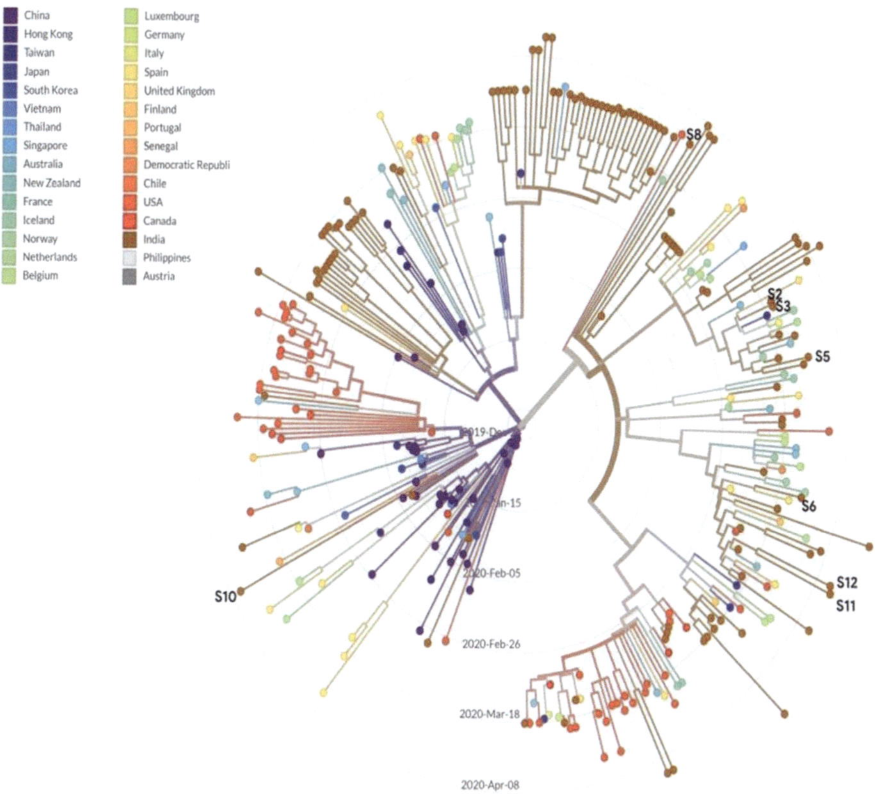

Fig. 3.2 Radial phylogenetic time-tree of the sequences from West Bengal, rest of India, and the world. The sequences from West Bengal included in this report are marked. (Adopted from Maitra et al. [7])

3.2 Lockdown and Pan-India Genome Sequencing Initiatives

Nonpharmaceutical reduction strategies were implemented in January 2020, which consisted of contact-tracing methods, thermal screening of international travelers, mandatory wearing of masks, and the shutdown of all educational and other public institutions across the country. In March 2020, India launched one of the largest testing databases in the world. On April 15, 2020, TRUENAT, the first point-of-care molecular diagnostic COVID-19 test, was approved by ICMR and WHO.

In context to pharmaceutical responses, on January 16, 2021, the government of India started the nationwide administration of COVID-19 vaccines. Within 1 year, >41.4 M vaccines were administered, and by October 31, 2022, it increased to >950 M. Oxford–Astrazeneca (developed under the license by the Serum Institute of India, under the trade name Covishield) and Covaxin (developed in India by Bharat Biotech) vaccines were approved for use. Restricted use of Sputnik V vaccine (produced by the Serum Institute of India) was permitted in emergencies by the Drugs Controller General of India on April 13, 2021.

The pandemic mitigation efforts were complemented by large-scale collaborative genome surveillance initiatives, which included the Pan-India 1000 SARS-CoV-2 genomic consortium, launched in 2020 [8], which involved numerous stakeholders and institutes associated with the Department of Biotechnology (DBT) and was coordinated by the National Institute of Biomedical Genomics (NIBMG) in West Bengal. Similarly, the major Pan-India network, also known as the Indian SARS-CoV-2 Genomics Consortium (INSACOG), was launched on December 28, 2020. This network involved the concerted efforts from the Ministry of Health and Family Welfare (MOHFW), DBT, Council for Scientific and Industrial Research (CSIR), Indian Council of Medical Research (ICMR), National Centre for Disease Control (NCDC), Central Surveillance Unit (CSU), and the Integrated Disease Surveillance Programme (IDSP). The activities of Pan-India 1000 SARS-CoV-2 RNA Genome Sequencing Consortium of DBT institutes resulted in the first systematic insight into the viral genome variants and evolution across the various parts of India. The data generated from these consortiums proved invaluable to monitor and detect the viral variants and to monitor the pandemic progression throughout the country ([7], biorxiv) (Fig. 3.3).

Through this study, we came to know that, in India, the various circulating SARS-CoV-2 lineages have arrived mainly from international travelers, arriving from USA, Europe, and East Asia (Fig. 3.4). A2a was the predominating lineage,

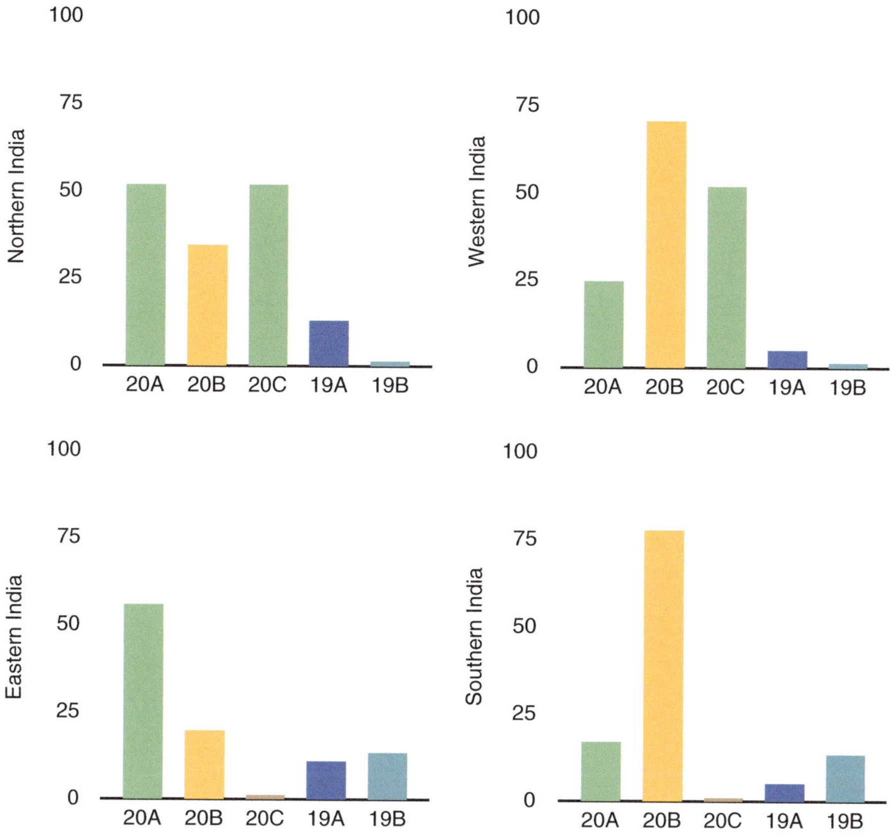

Fig. 3.3 SARS-CoV-2 haplotype distribution in the four regions of India. (Adopted from Maitra et al. [7])

with some other parental haplotypes like 19A/B. Among the other mutations, the D614G mutant type was found to be emerging in almost all parts of the country. The temporal haplotype diversity scenario appeared to be similar across India, peaking between March and May, but the various Indian states revealed different proportions of the haplotypes. Spatial and temporal patterns in the viral genome sequences reveal that some mutations showed a high frequency in one state as compared to the others. The in-depth study of the mutations across India dissected through this study had helped in a better disease management.

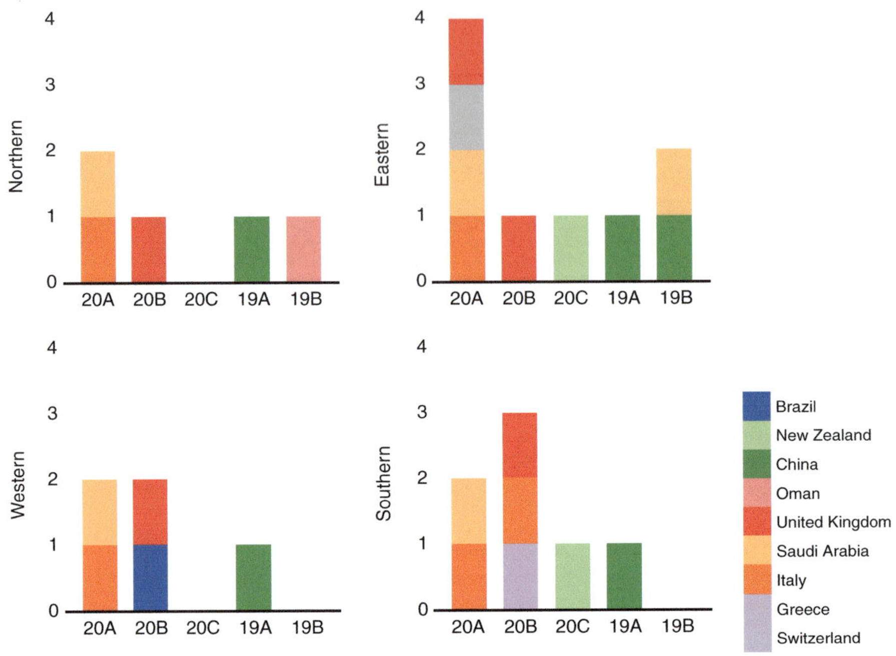

Fig. 3.4 Introduction of SARS-CoV-2 haplotypes by foreign countries in different Indian regions. (Adopted from Maitra et al. [7])

3.3 INSACOG

The foundation of Indian SARS-CoV-2 Genomics Consortium (INSACOG) was laid down in December 2020 with the primary aim to expand the SARS-CoV-2 whole genome sequencing initiatives across the nation (Fig. 3.5). The rapidly evolving virus requires constant monitoring to understand the sequence changes and correlate the mutations clinically to have complete information on immune escape, disease severity, transmissibility, etc. India's greatest network of genome sequencing was thus set up.

INSACOG was established with 10 national laboratories, and it expanded to 58 laboratories across the country in the Hub-and-spoke model, where the 10 national laboratories act as mentor organizations for the other laboratories which were added to the network, as part of infrastructure and genome sequencing capacity building efforts of the consortium.

The purpose of formulating INSACOG was to elucidate the spread and evolution of SARS-CoV-2, elucidate the genomic constellation, track the accumulated mutations along with the altered pathogenicity of the mutants, and understand if any local outbreaks occur due to the transmission of a specific viral lineage. The generated genomic data has direct implications for public health intervention strategies, development of diagnostics and variant-specific therapies, improving the understanding of re-infection cases and differentiate between prolonged infection and re-infection.

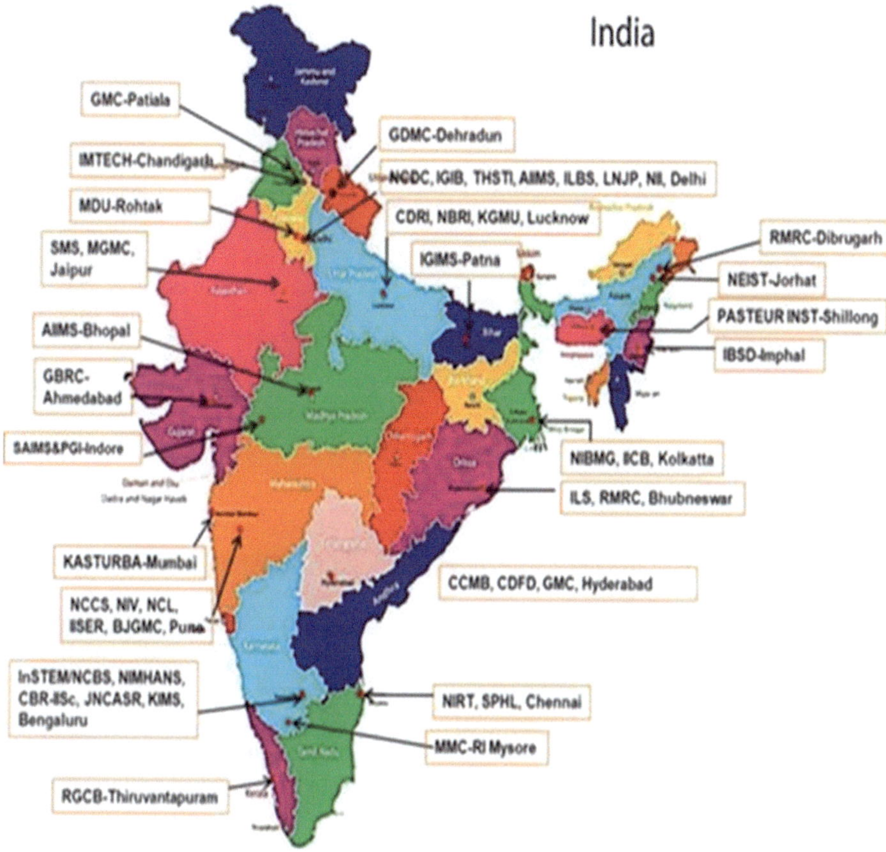

Fig. 3.5 Distribution of INSACOG laboratories across India

The sequenced viral genomes analyzed by the respective sequencing laboratories were regularly shared with the Central Surveillance Unit (CSU) working under Integrated Disease Surveillance Programme (IDSP) at the National Centre for Disease Control (NCDC), who further correlated these data with the field data trends and established the associations between the emerging SARS-CoV-2 variants and epidemiological drifts based on the COVID data generated by the State and District Surveillance Units of IDSP.

Over 3 lakh samples have been sequenced and analyzed by the INSACOG labs so far. Over 2.71 lakh sequences have been uploaded from India in the global database GISAD. In 2021, INSACOG identified a novel triple mutant from the outbreaks of Maharashtra, characterized by the mutations L452R, E484Q, and P681R. This mutant finally received a lineage B.1.617 (Delta). Soon, Delta sublineages, B.1.617.1, B.1.617.2, and B.1.617.3 started appearing upon the accumulation of subsequent mutations in the spike and other genes. INSACOG laboratories always closely monitored the emerging mutations and regularly reported to the

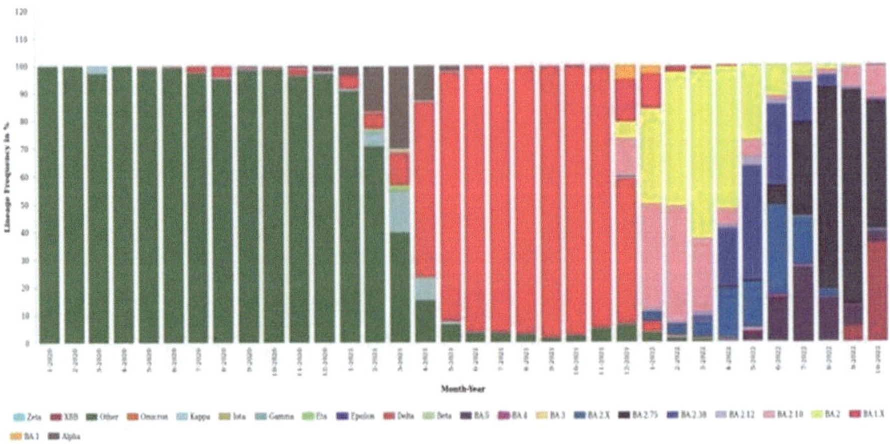

Fig. 3.6 Distribution of SARS-CoV-2 lineages (2020–22) in India

government. Henceforth, appropriate public health measures were undertaken, and the transmission of the variants was mitigated (Fig. 3.6).

The INSACOG consortium, in its efforts to curb the turnaround time and reporting, developed the Integrated Health Information Platform (IHIP), and through this, the data were shared by the sequencing laboratories with NCDC, as well as the states. The consortium had consolidated data at the Indian Biological Data Centre (IBDC) at RCB, Faridabad. The mirrored data repositories were also maintained at portals maintained by NIBMG, Kalyani. The first case of Omicron VoC was detected by INSACOG laboratory in Karnataka from an international traveler who arrived on November 28, 2021.

To gain detailed insights into the transmissibility and immune evasion potential of the Delta VoC of SARS-CoV-2, investigations were carried out by INSACOG and its collaborators, which were published in high-impact journals like Nature and Science. In one of such publications, the virus replication kinetics and spike-mediated infectivity assay showed possible elucidation of the dominance of the Delta (B.1.617.2) variant, which had higher replication potential and spike-mediated entry than other variants like Kappa (B.1.617.1) or Alpha (B.1.1.7) [9]. In another study, it was reported that the high transmission potential and partial immune evasion of SARS-CoV-2 Delta VoC was responsible for an overwhelming surge of the virus in various states like Delhi, Maharashtra, and Kerala [10].

Further, to enhance the sequencing capacity and facilitate the academia–industry partnership, and based on the acceptance of terms and conditions, the following three private laboratories were included based on their sequencing facility and trained capabilities, and the hub labs were assigned for them. (a) Strand Life Sciences Pvt., Ltd., Bengaluru (Hub lab: Instem); (b) Caringdx Mahajan Imaging Pvt., Ltd., Delhi (Hub lab: IGIB); and (c) Gene String Diagnostics, Delhi (Hub lab: IGIB).

3.4 Hospital Network

To establish the clinical correlation with the pathogen dynamics, INSACOG established a dedicated Hospital–IGSL network. The overall project was conceptualized as a longitudinal cohort study. Two approaches for enrolling the SARS-CoV-2-positive patients were followed. The first approach (Group 1) enrolled participants from testing centers, and the second approach (Group 2) enrolled hospitalized patients. Both the cohorts were followed up for immediate outcomes and post-COVID-17 sequelae and complications. Patients who report and test positive for SARS-CoV-2 infection at the testing centers were included and followed up for studying their clinical course and need for hospitalization. The hospitalized patients who were SARS-CoV-2-infected were studied for the risk factors for moderate to severe disease, clinical course, and outcomes. The exposure assessment and detailed clinical history were documented. INSACOG produced a weekly report on the status of SARS-CoV-2 genomic variants across the country, which was used by the Government of India (GOI) for strategic planning to control the pandemic. As the pandemic subsided, the frequency of reporting was made fortnightly. Besides the report for the use of the GOI, INSACOG also produced a weekly (and later fortnightly) bulletin for the information of the press and the public at large.

3.5 Evolution of SARS-CoV-2 Variants in India

Initially, a comprehensive study from ILS, Bhubaneswar, revealed that 66.6% SARS-CoV-2-infected symptomatic patients had elevated levels of lactate dehydrogenase, C-reactive protein, serum SGPT, and SGOT, along with ferritin [11]. Accumulation of the various mutations across the viral genes led to the fast evolution of SARS-CoV-2. During the detrimental second pandemic wave in India (April–May 2021), a study performed by the NIBMG group of researchers reported the whole genome sequencing of 239 SARS-CoV-2 RNA samples (collected from both the adult and pediatric patients), across Eastern India (West Bengal). This was the first report from eastern India which revealed a unique signature set of eight non-spike co-appearing mutations, which could have caused the evolution of a distinct Delta subcluster. 41.4% of the Delta strains harbored this array of co-appearing non-spike mutations. Continuous monitoring of such non-spike mutations will assist in the designing of future vaccines against those SARS-CoV-2 variants that might evade the immunity, produced by the current vaccines (Fig. 3.7) [12]. A study from THSTI, India portrayed that irrespective of the earlier acquired immunity, infections caused by SARS-CoV-2 Omicron variant was comparatively less severe than the infections caused by Delta VoC [13]. Another study from the same institute represented a proof of principle for the formulation of therapeutics to curb the severe airway inflammation caused by the inflammatory pathway triggered by host cytokines like IL-9 in the case of SARS-CoV-2 infection [14].

The phylodynamics of different SARS-CoV-2 variants across the consecutive waves of COVID-19 infection were studied in various states of India in a study

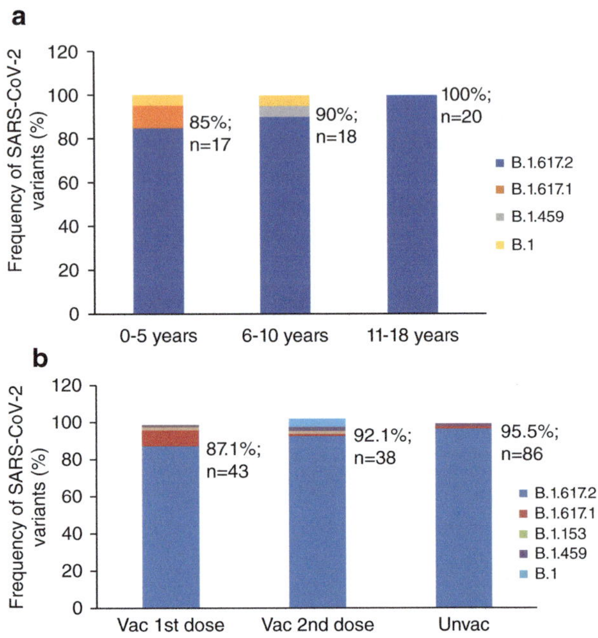

Fig. 3.7 Frequency of the SARS-CoV-2 Delta variant during the second pandemic wave across eastern India among (**a**) pediatric age group (below 18 years) and (**b**) adult patients (above 18 years). (Adopted from Banerjee et al. [12])

conducted at NIBMG. Such data from a geographically important state like West Bengal (East) was compared with the scenario from other states like Delhi (North), Maharashtra (West), and Karnataka (South). The authors have identified 57 SARS-CoV-2 pango-lineages among the 3198 viral genomes, with alteration in their predominance and distribution, as well as contrasting constellation of mutations in different viral proteins, across different pandemic waves. The study portrayed a downward trend in mutation rate in specific viral genes across the state of West Bengal across the consecutives waves, while a simultaneous surge in infection counts across the four states. Toward the end of the second pandemic wave, the viral phylogenetic diversity reduced, while majorly one single lineage, B.1.617.2 (Delta), predominated across all the four states (Table 3.1) [15].

A substantial reduction in the neutralizing ability of both vaccine-induced and vaccine combined with SARS-CoV-2 infection-induced antibodies was observed in a study conducted at THSTI with the omicron VoC, and this study elucidated the role of Omicron in host-immune evasion (Medigeshi et.al, 2022). Importantly, a group from IISc, Bangalore, has developed a SARS-CoV-2 virus-like particle (VLP), a platform to evaluate the impact of the emerging mutations in different variants of concern (VoC). Specific mutations have been incorporated in the VLP, and alterations were investigated in the context of viral entry, immunogenicity, and pathogenicity [16].

Table 3.1 Top lineages for each state having more than or equal to 5% frequency in each wave category

State	Pre-wave 1	Wave 1	Pre-wave 2	Wave 2
West Bengal	*B.1 (n = 123; 64.06%)*, B.1.540 (*n* = 17; 8.85%), A.9 (*n* = 15; 7.81%)	*B.1 (n = 112; 24.3%)*, *B.1.36 (n = 81; 17.57%)*, *B.1.1.216 (n = 67; 14.53%)*, *B.1.1 (n = 52; 11.28%)*, B.1.1.220 (*n* = 31; 6.72%), B.1.540 (*n* = 26; 5.64%)	*B.1.618 (n = 121; 20.27%)*, B.1.617.1 (*n* = 80; 13.4%), B.1.1.216 (n = 77; 12.9%), *B.1.1.7 (n = 69; 11.56%)*, B.1.1.220 (n = 49; 8.21%), *B.1.36 (n = 47; 7.87%)*, *B.1.1 (n = 33; 5.53%)*	*B.1.617.2 (n = 1294; 66.43%)*, *B.1.617.1 (n = 510; 26.18%)*
Delhi	*B.6 (n = 149; 45.43%)*, B.1.36 (*n* = 69; 21.04%), B.6.6 (*n* = 56; 17.07%), *B.1.1 (n = 24; 7.32%)*	*B.1.36 (n = 440; 45.83%)*, *B.1 (n = 172; 17.92%)*, *B.1.1.216 (n = 135; 14.06%)*, *B.1.1 (n = 55; 5.73%)*	*B.1.1.7 (n = 193; 32.38%)*, *B.1 (n = 118; 19.8%)*, *B.1.617.1 (n = 59; 9.9%)*, *B.1.36 (n = 55; 9.23%)*, B.1.525 (n = 33; 5.54%)	*B.1.617.2 (n = 604; 48.67%)*, *B.1.1.7 (n = 311; 25.06%)*, *B.1.617.1 (n = 172; 13.86%)*
Karnataka	B.1.1.46 (*n* = 51; 23.61%), *B.6 (n = 50; 23.15%)*, *B.1.210 (n = 27; 12.5%)*, *B.1.1 (n = 23; 10.65%)*, *B.1 (n = 21; 9.72%)*, B.1.36.8 (*n* = 11; 5.09%)	B.1.36.29 (*n* = 80; 29.74%), *B.1.1 (n = 44; 16.36%)*, B.1.1.46 (*n* = 22; 8.18%), B.1.560 (*n* = 22; 8.18%), *B.1.36 (n = 18; 6.69%)*, *B.1 (n = 18; 6.69%)*, B.1.1.7 (*n* = 15; 5.58%)	*B.1.36.29 (n = 50; 23.81%)*, *B.1.1.7 (n = 31; 14.76%)*, *B.1.617.1 (n = 23; 10.95%)*, B.1.560 (*n* = 17; 8.1%), B.1.617.2 (*n* = 17; 8.1%), *B.1 (n = 13; 6.19%)*, *B.1.1 (n = 12; 5.71%)*	*B.1.617.2 (n = 733; 72.15%)*, *B.1.617.1 (n = 145; 14.27%)*, *B.1.1.7 (n = 75; 7.38%)*

(continued)

Table 3.1 (continued)

State	Pre-wave 1	Wave 1	Pre-wave 2	Wave 2
Maharashtra	B.1.1.306 (*n* = 403; 58.75%), *B.1.210 (n = 117; 17.06%)*, *B.1 (n = 54; 7.87%)*, *B.6 (n = 38; 5.54%)*	B.1.1.306 (*n* = 484; 61.81%), B.1.210 (*n* = 103; 13.15%), *B.1 (n = 40; 5.11%)*	*B.1.617.1 (n = 193; 29.42%)*, *B.1.36.29 (n = 109; 16.62%)*, B.1.1.306 (*n* = 71; 10.82%), *B.1.36 (n = 66; 10.06%)*, *B.1.1 (n = 34; 5.18%)*, *B.1.1.7 (n = 33; 5.03%)*	*B.1.617.1 (n = 1563; 44.18%)*, *B.1.617.2 (n = 1206; 34.09%)*

Adopted from Singh et al. [15]

3.6 Lessons Learned, Achievements, and Way Forward

The study of the biology of the disease progression and transmission is important to take proper decisions on the future preventive measures, prediction of the vaccine of choice, and health policies. Confirmed decisions can be taken about international travel between countries. India has combatted well against the pandemic and enabled the viral genomes of patients to be analyzed quickly with the application of standardized pipelines. Focus has been given to the analysis of SARS-CoV-2 transmission dynamics within the hotspots of local transmission.

The population density, hygiene habits, and public health policies across the geographical regions in India are unique in each case, and Indian scientific communities have addressed the viral disease burden with respect to these unique variants extremely well. A particular variant of concern can lead to the upsurge of cases in the future and infiltration of subsequent waves of this pandemic. Hence, continuous preparedness for future pandemics along with the ongoing sewage surveillance programs would be helpful to quickly adopt preventive measures depending on the situation.

SARS-CoV-2 is never the ultimate virus we are fighting the detrimental war against. The continuously evolving virus will be a continued focus for the formulation of the scientific strategies to address the threats posed against human health. This ecosystem which we have built was put to a great challenge during the pandemic.

This publication brings out the complete story of how India strongly fought the pandemic with a well-articulated timely scientific response. It documents the stories from the key players and bring out not only the success but also the challenges and how innovative solutions were found in the shortest time frame to address these. This publication is a comprehensive documentation of how science fought the pandemic and we won the war against the single strand.

Competing Interest The authors have no conflicts of interest to declare that are relevant to the content of this chapter.

References

1. Chavda VP, Vuppu S, Mishra T, Kamaraj S, Patel AB, Sharma N, Chen ZS. Recent review of COVID-19 management: diagnosis, treatment, and vaccination. Pharmacol Rep. 2022;74(6):1120–48. https://doi.org/10.1007/s43440-022-00425-5. Epub 2022 Oct 10. PMID: 36214969; PMCID: PMC9549062.
2. Kaur J, Kaur J, Dhama AS, Kumar V, Singh H. Management of COVID-19 pandemic data in India: Challenges faced and lessons learnt. Front Big Data. 2021;4:790158. https://doi.org/10.3389/fdata.2021.790158. PMID: 34917935; PMCID: PMC8670559.
3. Yang H, Rao Z. Structural biology of SARS-CoV-2 and implications for therapeutic development. Nat Rev Microbiol. 2021;19(11):685-700. https://doi.org/10.1038/s41579-021-00630-8. Epub 2021 Sep 17. PMID: 34535791; PMCID: PMC8447893.
4. Arya R, Kumari S, Pandey B, Mistry H, Bihani SC, Das A, Prashar V, Gupta GD, Panicker L, Kumar M. Structural insights into SARS-CoV-2 proteins. J Mol Biol. 2021;433(2):166725.

https://doi.org/10.1016/j.jmb.2020.11.024. Epub 2020 Nov 24. PMID: 33245961; PMCID: PMC7685130.

5. Harvey WT, Carabelli AM, Jackson B, Gupta RK, Thomson EC, Harrison EM, Ludden C, Reeve R, Rambaut A; COVID-19 Genomics UK (COG-UK) Consortium; Peacock SJ, Robertson DL. SARS-CoV-2 variants, spike mutations and immune escape. Nat Rev Microbiol. 2021;19(7):409–24. https://doi.org/10.1038/s41579-021-00573-0. Epub 2021 Jun 1. PMID: 34075212; PMCID: PMC8167834.
6. Azhar M, Phutela R, Kumar M, Ansari AH, Rauthan R, Gulati S, Sharma N, Sinha D, Sharma S, Singh S, Acharya S, Sarkar S, Paul D, Kathpalia P, Aich M, Sehgal P, Ranjan G, Bhoyar RC; Indian CoV2 Genomics; Genetic Epidemiology (IndiCovGEN) Consortium; Singhal K, Lad H, Patra PK, Makharia G, Chandak GR, Pesala B, Chakraborty D, Maiti S. Rapid and accurate nucleobase detection using FnCas9 and its application in COVID-19 diagnosis. Biosens Bioelectron. 2021;183:113207. https://doi.org/10.1016/j.bios.2021.113207. Epub 2021 Apr 5. PMID: 33866136; PMCID: PMC8020606.
7. Maitra A, Sarkar MC, Raheja H, Biswas NK, Chakraborti S, Singh AK, Ghosh S, Sarkar S, Patra S, Mondal RK, Ghosh T, Chatterjee A, Banu H, Majumdar A, Chinnaswamy S, Srinivasan N, Dutta S, DAS S. Mutations in SARS-CoV-2 viral RNA identified in Eastern India: Possible implications for the ongoing outbreak in India and impact on viral structure and host susceptibility. J Biosci. 2020;45(1):76. https://doi.org/10.1007/s12038-020-00046-1. PMID: 32515358; PMCID: PMC7269891.
8. Sauche YS, Seshasayee ASN, Palakodeti D, Bashyam MD, Parida A, Das S. PAN-INDIA 1000 SARS-CoV-2 RNA genome sequencing reveals important insights into the outbreak. BioRxiv. 2020; https://doi.org/10.1101/2020.08.03.233718.
9. Mlcochova P, Kemp SA, Dhar MS, Papa G, Meng B, Ferreira IATM, Datir R, Collier DA, Albecka A, Singh S, Pandey R, Brown J, Zhou J, Goonawardane N, Mishra S, Whittaker C, Mellan T, Marwal R, Datta M, Sengupta S, Ponnusamy K, Radhakrishnan VS, Abdullahi A, Charles O, Chattopadhyay P, Devi P, Caputo D, Peacock T, Wattal C, Goel N, Satwik A, Vaishya R, Agarwal M; Indian SARS-CoV-2 Genomics Consortium (INSACOG); Genotype to Phenotype Japan (G2P-Japan) Consortium; CITIID-NIHR BioResource COVID-19 Collaboration; Mavousian A, Lee JH, Bassi J, Silacci-Fegni C, Saliba C, Pinto D, Irie T, Yoshida I, Hamilton WL, Sato K, Bhatt S, Flaxman S, James LC, Corti D, Piccoli L, Barclay WS, Rakshit P, Agrawal A, Gupta RK. SARS-CoV-2 B.1.617.2 Delta variant replication and immune evasion. Nature. 2021;599(7883):114–9. https://doi.org/10.1038/s41586-021-03944-y. Epub 2021 Sep 6. PMID: 34488225; PMCID: PMC8566220.
10. Dhar MS, Marwal R, Vis R, Ponnusamy K, Jolly B, Bhoyar RC, Sardana V, Naushin S, Rophina M, Mellan TA, Mishra S, Whittaker C, Fatihi S, Datta M, Singh P, Sharma U, Ujjainiya R, Bhatheja N, Divakar MK, Singh MK, Imran M, Senthivel V, Maurya R, Jha N, Mehta P, A V, Sharma P, Vr A, Chaudhary U, Soni N, Thukral L, Flaxman S, Bhatt S, Pandey R, Dash D, Faruq M, Lall H, Gogia H, Madan P, Kulkarni S, Chauhan H, Sengupta S, Kabra S; Indian SARS-CoV-2 Genomics Consortium (INSACOG); Gupta RK, Singh SK, Agrawal A, Rakshit P, Nandicoori V, Tallapaka KB, Sowpati DT, Thangaraj K, Bashyam MD, Dalal A, Sivasubbu S, Scaria V, Parida A, Raghav SK, Prasad P, Sarin A, Mayor S, Ramakrishnan U, Palakodeti D, Seshasayee ASN, Bhat M, Shouche Y, Pillai A, Dikid T, Das S, Maitra A, Chinnaswamy S, Biswas NK, Desai AS, Pattabiraman C, Manjunatha MV, Mani RS, Arunachal Udupi G, Abraham P, Atul PV, Cherian SS. Genomic characterization and epidemiology of an emerging SARS-CoV-2 variant in Delhi, India. Science. 2021;374(6570):995–9. https://doi.org/10.1126/science.abj9932. Epub 2021 Oct 14. PMID: 34648303; PMCID: PMC7612010.
11. Chatterjee S, Datey A, Sengupta S, Ghosh A, Jha A, Walia S, Singh S, Suranjika S, Bhattacharya G, Laha E, Keshry SS, Ray A, Pani SS, Suryawanshi AR, Dash R, Senapati S, Beuria TK, Syed GH, Prasad P, Raghav SK, Devadas S, Swain RK, Chattopadhyay S, Parida A. Clinical, virological, immunological, and genomic characterization of asymptomatic and symptomatic cases with SARS-CoV-2 infection in India. Front Cell Infect Microbiol. 2021;11:725035. https://doi.org/10.3389/fcimb.2021.725035. PMID: 34993157; PMCID: PMC8724424.

12. Banerjee A, Mazumder A, Roy J, Das J, Majumdar A, Chatterjee A, Biswas NK, Chawla Sarkar M, Das S, Dutta S, Maitra A. Emergence of a unique SARS-CoV-2 Delta sub-cluster harboring a constellation of co-appearing non-Spike mutations. J Med Virol. 2023;95(1):e28413. https://doi.org/10.1002/jmv.28413. PMID: 36541745; PMCID: PMC9878222.
13. Thiruvengadam R, Rizvi ZA, Raghavan S, Murugesan DR, Gosain M, Dandotiya J, Ayushi SS, Pandey AK, Wadhwa N, Bhatnagar S, Awasthi A, Garg PK. Clinical and experimental evidence suggest omicron variant of SARS-CoV-2 is inherently less pathogenic than delta variant independent of previous immunity. Eur J Med Res. 2023;28(1):421. https://doi.org/10.1186/s40001-023-01373-3. PMID: 37821945; PMCID: PMC10566023.
14. Sadhu S, Dalal R, Dandotiya J, Binayke A, Singh V, Tripathy MR, Das V, Goswami S, Kumar S, Rizvi ZA, Awasthi A. IL-9 aggravates SARS-CoV-2 infection and exacerbates associated airway inflammation. Nat Commun. 2023;14(1):4060. https://doi.org/10.1038/s41467-023-39815-5. PMID: 37429848; PMCID: PMC10333319.
15. Singh AK, Laskar R, Banerjee A, Mondal RK, Gupta B, Deb S, Dutta S, Patra S, Ghosh T, Sarkar S, Ghosh S, Bhattacharya S, Roy D, Chakraborty A, Chowdhury M, Mahaptra S, Paul A, Mazumder A, Chowdhury A, Chatterjee SS, Sarkar A, Ray R, Pal K, Jana A, Barik G, Ganguly S, Chatterjee M, Majhi D, Bandopadhyay B, Das S, Maitra A, Biswas NK. Contrasting distribution of SARS-CoV-2 lineages across multiple rounds of pandemic waves in West Bengal, the gateway of east and north-east states of India. Microbiol Spectr. 2022;10(4):e0091422. https://doi.org/10.1128/spectrum.00914-22. Epub 2022 Jul 19. PMID: 35852336; PMCID: PMC9430150.
16. Raheja H, Das S, Banerjee A, Dikshaya P, Deepika, Mukhopadhyay D, Ramachandra SG, Das S. RG203KR mutations in SARS-CoV-2 Nucleocapsid: assessing the impact using Virus-like particle model system. Microbiol Spectr. 2022;10(4):e0078122. https://doi.org/10.1128/spectrum.00781-22.

Leveraging the Scientific Strength to Test, Track, and Treat: A Public Health Perspective

4

Sujeet Singh

SARS, which is severe acute respiratory syndrome and affects the lower respiratory tract epithelial cells, is caused by a single-stranded RNA virus of the coronaviridae family. It not only affects the lungs but also impacts the brain neurons, mucosal cells of the intestines, tubular epithelial cells of the kidneys, and brain neurons. It was first reported in 2002 and spread as SARS in 2003. One of the new virus of this family which originated again from Wuhan, China, as per reports from the wild animals, in December 2019 named SARs-CoV is a virus which has an envelope of single-stranded RNA, and this was responsible for the COVID-19 pandemic.

The first official case of COVID-19 was recorded on December 31, 2019, when the World Health Organization (WHO) was informed of cases of pneumonia in Wuhan, China, with no known cause. Emergency meetings were convened on the new year day to understand the disease. WHO issued an alert on 43 cases. It was a matter of concern for public health. Alarm bells started anticipating something serious, something big, with the only information on transmission being the respiratory mode, from the human–bird/seafood interface. On January 7, the Chinese authorities identified this novel coronavirus, and initially named it as SARs-nCoV-2019, as the cause of these cases. Based on earlier experience with SARS and MERS and the established modes of transmission of respiratory viruses, infection, and prevention control, guidance documents were developed primarily for the surveillance of incoming travelers and linking it with in-country surveillance by the IDSP network including those to protect health workers from droplet and contact precautions and airborne precautions for aerosol- generating procedures conducted by health workers. After the detection of the first case outside China on January 13, Covid −19 had spread to over 32+ countries outside China, including Japan, S. Korea, Australia,

S. Singh (✉)
Public Health Specialist, Former Director and Advisor, National Centre for Disease Control, Ministry of Health & Family Welfare, Govt of India, New Delhi, India
e-mail: sujeet647@gmail.com

© The Editor(s) (if applicable) and The Author(s), under exclusive license to Springer Nature Singapore Pte Ltd. 2025
R. Swarup et al. (eds.), *At War with the Single Strand*,
https://doi.org/10.1007/978-981-96-8122-8_4

Singapore, Spain, and USA. Italy (with over 1000 cases) and Iran (530 cases) became hotspots in February 2020 itself.

The spread of the infectious disease was linked with a number of complex factors of evolution of the pathogen, environmental changes, human behavioral aspects, and the human–animal interactions. The appearance of new viruses resulting in new infectious diseases and reappearance of infectious diseases are generally related to the dynamic processes which drive changes in community structure and their interface with the pathogen.

A highly infectious SARS infection detected during 2003, due to the emergence of the new corona virus, emerged possibly due to the crossover of viruses causing infection among fruit bats, spread within weeks, starting from Hong Kong and then spread across 37 countries in a very short period. Of the approximately 8000 worldwide cases that occurred during this pandemic, about 25% of patients required mechanical ventilation in the ICU, and about 10% of infected patients died.

SARS has flulike symptoms of fever, chills, cough, and malaise, along with frequent shortness of breath. A common cause of death during this pandemic was diffuse alveolar damage (DAD). In addition, SARS typically compromises the immune response, which increases lung injury.

Rapid transmission with Ro about 4, high infectivity, significant morbidity, and mortality during SARS led countries adopt various known public health interventions which included screening of incoming travelers at airports to detect suspects, subjecting travelers having a history of travel to affected countries or contact with a case to testing, and management of confirmed cases in dedicated facilities. However, the scientific evidence of the above interventions were inconclusive. The 2003–2004 SARS pandemic showed that a prompt, coordinated worldwide response could help contain the disease. Although SARS rapidly spread across the world through international air travelers, the virus itself was not transmitted through air. Thus adherence to the basic infection control practice of thorough handwashing was stressed upon, since this practice was implemented with precautions for the control of droplet, the 2003 SARS infection ultimately subsided in 6 months. Public health measures like screening for detection of suspects among international travelers arriving from affected countries, quarantining, and testing of suspects were also subsequently adopted based on the experiences of SARS during H1N1, EVD, and MERS-CoV.

Hence, globally, the initial measures planned during Covid 19 were based on the known modes of transmission (direct and/or respiratory droplets) of respiratory viruses. There were no scientific evidence or information from credible sources on the virus sources/modes of transmission of disease agents and infectivity period to plan preventive and control strategies. There were a lot of rumors floating, speculations and scenarios like fomites or articles getting infected, ways in which virus can make articles affected, duration of survival of the virus outside the human body, modes/chemicals for disinfection, etc. In addition to nonavailability of diagnostic kits, primers and probes added to the confusion even in the minds of experts.

The response of Ministry of Health and Family Welfare to COVID 19 outbreak (which turned to global pandemic) was initially driven by the National Centre for

Disease Control, the premier public health institute of the country, which coordinates surveillance and response of outbreak-prone diseases in the country through IDSP network and monitors the implementation of public health measures. The detection of disease agents or events which requires advanced diagnostic tests or kits or development of primers/probes is primarily done by the National Institute of Virology (NIV), Pune, under Indian Council of Medical Research (ICMR). During initial days, only strategic planning was targeted to avoid the entry of infected persons/agents, based on disease scenarios (of fast spreading, high mortality, limited impact of public health measures, no specific drugs or management line) in China, S. Korea, Japan, and other countries. Airborne precautions for aerosol-generating procedures were conducted by health workers. IDSP states that units were alerted on global and country situation.

China on January 13, 2020 publicly shared the genetic sequence of COVID-19.

India's leading scientific body ICMR and NCDC internally coordinated between microbiology and biotechnology and through assistance from WHO and CDC developed indigenous testing capacities in the next 10 days. The labs focused primarily on testing through RT-PCR and also initiated limited serological and sequencing studies for the RNA virus. Screening of international travelers was initiated for the flights coming from China as early as January 17, 2020. On January 30, 2020, the WHO DG declared the COVID-19 outbreak a public health emergency of international concern (PHEIC) .

The joint monitoring group under DGHS asked NCDC to prepare the guidelines and strategies for the surveillance of incoming travelers. However, very soon, information on more cases in other areas, in spite of full sanitization of seafood market, started coming in, and it was a clear pointer to other modes of disease spread. The joint monitoring group that consisted of multidimensional experts planned the screening strategy, quarantine, and isolation guidelines, including tools primarily for the surveillance of incoming travelers and linking point of entry screening with IDSP for in-country community surveillance by the IDSP network from droplet and contact precautions. Airport screening initiated on January 23, 2020, was linked to community follow-up by IDSP of travelers arriving from China (Wuhan).

Considering the importance of the laid down epidemiological principles, specific plans were developed for active case search among persons having the history of international travel, detecting suspects, tracking and testing the quarantined contacts, and management of cases in isolation wards/health facilities. Subsequently, further spread from small clusters was limited by the creation of containment zones. The aim was to isolate the infected/exposed persons in a dedicated facility for 14 days till the person becomes noninfectious. Kerala additionally quarantined all the arriving travelers for 14 days, to watch for any infection among the travelers (who might have been in incubation phase during the airport screening). This led to tracking of first three cases on January 30 and February 2 and 3 among Wuhan returnee students. Immediately, the IDSP network was activated in around 12 states, for tracing and quarantining the in-flight contacts of three positive cases. The result was marvelous; no further cases were detected in the following 28+ days.

The country aggressively continued surveillance through airport screening and community tracking and quarantining through HCW at PoEs and through IDSP in states in an effort to detect the infected persons either through screening at the points of entry and, subsequently, through surveillance in community for those who could not be detected at the points of entry due to being asymptomatic or in incubation phase. These efforts reported first cases on March 2, 2020 (after successfully tackling January cases), 1 in Delhi, 1 in Telangana, and 1 in Jaipur (all among travelers returning from affected countries). Epidemiological concerns such as transmission dynamics, modes of transmission (via environment and articles), infectivity period, virus survival outside infected persons among many others had left public health specialist plan strategies taking a precautious approach. The scientific data-reliable information from credible experts was a big felt need. In such a scenario, NCDC, the country's leading public health institution, was assessing the global epidemiological scenario, secondary data from affected and unaffected, including various surveillance and response strategies. The top epidemiologists were engaged in scientific discussions, while many countries in the world were already closing the international borders and putting domestic community in partial or total lockdown.

The WHO declared COVID-19 as a pandemic on March 11, 2020, and asked all countries to take urgent and aggressive action against this public health crisis.

Governance Structures Steering the pandemic surveillance and response under the overall leadership of the PM of India, the whole-of-government structure with focused comprehensive attention on all aspects was put in place, which was effectively coordinated with state-level governance structures going down right up to the district administrative level. Disaster management act was evoked which provided a legal framework for the implementation of the strategic plans from the national to state level.

Meetings of the Chief Ministers of all States and UT administrations and all stakeholders with the Honorable Prime Minister were held regularly to understand the issues and develop collaboration modalities with the states for effective COVID management. A group of ministers under the chairmanship of Health and Family Welfare Minister Dr. Harsh Vardhan, with Minister of External Affairs, Minister of Civil Aviation and Minister of State for Home Affairs, Minister of State for Shipping and Minister of State for Health and Family Welfare, reviewed the COVID situation with key technical and administrative officials of the Ministry of Health like ICMR, NCDC, and NITI Aayog. This committee met 22 times since its inception on February 3, 2020. A committee of secretaries under Cabinet Secretary also regularly reviewed with all related Ministries of Health, Defence, Ministry of External Affairs, Civil Aviation, Home, Biotechnology, Textiles, Pharma, Commerce, and other officials including State Chief Secretaries.

The Government of India constituted 11 empowered groups on March 29, 2020, on different aspects of COVID-19 management in the country to take informed decisions on issues ranging from (i) medical emergency planning, (ii) availability of hospitals, isolation and quarantine facility, disease surveillance, and testing, (iii) ensuring availability of essential medical equipment, (iv) augmenting human resource and capacity building, (v) supply chain and logistics management, (vi) coordination with private sector, (vii) economic and welfare measures, (viii) information, communication, and public awareness, (ix) technology and data management, (x) public grievance, and (xi) strategic issues related to lockdown. These groups were restructured on September 10, 2021 based on the need and evolving scenario.

The Health Ministry held videoconferences with the states regularly. The Joint Monitoring Group (JMG) under the Chairmanship of DGHS which advises MoHFW on technical matters met several times to assess the risk, review the preparedness and response mechanisms, and finalize technical guidelines.

The cluster containment was gradually expanded to large outbreaks from March 2 to April 4, 2020, respectively. The containment plans envisaged a strategy of breaking the chain of transmission by (i) defining the containment and buffer zones, (ii) applying strict perimeter control, (iii) intensive active house-to- house search for cases and contacts, (iv) isolation and testing of suspect cases and high-risk contacts, (v) quarantine of high-risk contacts, (vi) intensive risk communication to raise community awareness on simple preventive measures and need for prompt treatment-seeking, and (vii) strengthening of passive influenza-like illness (ILI)/severe acute respiratory illness (SARI) surveillance in the containment and buffer zones.

The laboratory network was strengthened to support the detection of infected cases in the shortest possible time, so that the infection from the community be detected at the earliest by developing new testing centers in district hospitals, medical colleges, and even private sector labs across various states. This helped reducing the infection burden in the community by shifting the asymptomatic to quarantine facilities and symptomatic to dedicated covid care centers with isolation facilities.

Although all the states and districts were effectively implementing the public health interventions, the respiratory mode of spread, which was evident during the asymptomatic phase, saw an exponential rise in cases from less than a 100 on March 15 to over 1000 by March end. The gradual worsening of the global situation, rising mortalities, lack of scientific knowledge, limited PPEs, treatment options, and overburdening of health facilities led to the decision of complete country-wide lockdown. India enforced 68 days of four-phased lockdown starting from March 24 to May 31, 2020, to deal with COVID-19, a much needed respite for firming up the all-round preparedness in surveillance and response structures right up to the district level, including the availability of PPEs and other logistics and provision of necessary funds. The increasing country-wide magnitude of this outbreak called for a pre-emptive, proactive, graded, whole-of-government, whole-of-society approach, built around a comprehensive strategy to prevent infections, save lives, and minimize impact.

From the situation of a single laboratory equipped to undertake testing for COVID in January 2020, there were 2288 laboratories conducting COVID-19 testing by December 2020. Laboratories were established in difficult terrains like Ladakh, Sikkim, Arunachal Pradesh, Nagaland, as well as other North Eastern states, such as Lakshadweep and Andaman and Nicobar Islands. India enhanced the testing, which crossed 1.5 million tests a day, much higher than that stipulated by WHO (140 tests per million population per day). Indigenous manufacturers of laboratory diagnostics or testing machines for COVID supported an indigenous production capacity of more than 10 lakh kits/day by the end of 2020.

A three-tier arrangement of health facilities was created for the appropriate management of COVID-19 cases: (i) COVID Care Center with isolation beds for mild or presymptomatic cases; (ii) Dedicated COVID Health Center (DCHC) with oxygen-supported isolation beds for moderate cases, and (iii) Dedicated COVID Hospital (DCH) with ICU beds for severe cases. Tertiary care hospitals under ESIC, Defence, Railways, paramilitary forces, Steel Ministry, etc. have been leveraged for case management.

As on December 21, 2020, 222 countries/territories worldwide were affected.

India's endeavor to manage COVID-19 through a whole-of-government and whole-of-society approach was able to limit its cases and deaths. As on December 29, 2020, a total of 10,224,303 confirmed cases (with 2,68,581 active cases comprising 2.62% of the total cases) were reported in India. 98,07,569(95.%92) cases had recovered, while the fatality cases stood at 1.45%, one of the lowest globally.

On one hand, countries were struggling to fight the COVID 19 pandemic to reduce the spread in new areas, using early quarantine and isolation, use of PPEs, and CAB to reduce mortality; on the other front, efforts were on to assess the extent of exposure immunity through sero-surveys, development of vaccines, trials of newer drugs, and use of home- and country-prevalent non-allopathic preparations to boost individual immunity. Some epidemiologists were also engaged in assessing the possibility of development of herd immunity, which in turn would provide community-level protection. ICMR/NCDC planned a country-wide sero-survey and, however, assessed very low levels of exposures with state-specific variations. NCDC epidemiologists coordinated and supported Ser-surveys in cities like MP, Rajasthan, Gujarat, UP, etc., that were regularly planned by states. However, some epidemiologists and microbiologists, understanding the behavior of RNA virus mutations in sequencing studies, for tracking the linkage of spread to mutations, had doubts. None of the predictions, modeling forecasts were able to provide any clues related to the future trends of the pandemic. The scientific community across many countries succeeded in developing candidate vaccines. Indian government-supported labs also developed vaccines against COVID 19.

This global situation got further complicated by significant mutations noticed during sequencing. Primarily, sequencing was only done in small number of RT-PCR-positive cases, directed to track the linking of affected areas and pointing to the routes of spread.

There was a sudden fourfold spurt in cases in Kent, northern England, where the exposure immunity was assessed to be fairly high during October and November 2020, and detection of the new Alpha variant led to a looming threat of emergence of newer variants of SARS-CoV 2. The Ministry of Health and Department of Biotechnology immediately coordinated the sequencing labs, working in silos primarily for research under different ministries, to establish INSACOG, a consortium of labs from NCDC, ICMR, CSIR, DBT, and DST with a focused whole genome of SARS-CoV 2 from targeted RT-PCR-positive samples to continuously track mutations/variants and made NCDC as the nodal center to assess the VOCs after the community impact on transmission dynamics and severity. The scientific coordination with global expertise on whole-genome sequencing strategy along with strengthened surveillance in the community assessed the impact of new variants on transmission, disease severity, mortality, and immune escape phenomenon. This helped to review and prioritize the control interventions for ensuring that disease spread is gradual and infected persons are promptly shifted for clinical management.

India also rolled out vaccinations targeting frontline workers, elderly, and those with comorbidities. However, availability of vaccine, logistics for delivery, coverage with two doses, and development of immunity in substantial populations were challenges for administration and health systems in various states. The complexities of mutations, particularly in the spike region of single-stranded RNA virus, led to the emergence of a new variant with double mutation (later named as Delta variant) in the later part of January, 2021 in the Vidarbha region of Maharashtra, detected on the basis of sudden increased (fourfold) hospitalization. The significant involvement of lungs and increased demand of oxygen led to high mortality among cases due to the Delta variant. The high transmission rate led to rapid spread to nearly all regions of the country.

By September 2021, dedicated vaccination efforts along with substantial exposure during the Delta wave led to the development of protective immunity (exposure and/or vaccine-induced). These substantially complemented public health interventions. As a planned strategy, close monitoring was done on new emerging variants due to recombinants and sublineages of variants through INSACOG sequencing labs and integrated surveillance system (IDSP) under the expert guidance of NCDC. Detection of the Omicron variant was done with more than 50 mutations, with around 30 mutations in the spike protein region of the virus. This led to a substantial rise in the rate of transmission across the countries including India. Effective coordination between INSACOG labs, IDSP network, and surveillance of hospitalization was done to promptly assess the public health impact of Omicron and its sublineages in the community, both among the nonvaccinated and vaccinated persons. Fortunately, no rise in severity or mortality was seen to be associated with Omicron or any of its sublineages in India. The rise in transmission was effectively intercepted with public health interventions and enforcement of covid-appropriate behavior (CAB). With the all-round gradual and sustained decline in active cases, hospitalization, and mortality, the surveillance strategy was focused on testing all ARI and ILI cases seeking hospitalization to keep a close watch on the trend of covid cases and infection in the community. By the end of September 2022, cases came down to below 5000, and the disease became endemic in most states of the country, with life returning to normal. Thereafter, a small peak was observed in April 2023, with new cases going up to 11,000 and then returning to less than 1000 by the month end.

Although many new variants like BF 5 were detected by the INSACOG lab network and followed up by NCDC, no significant public health impact could be seen in terms of rise in transmission or hospitalization.

4.1 Key Challenges: Technical

Key challenges faced and measures taken are listed in Table 4.1. Some key issues which needed special attention are as follows:

- High infection of HCWs and growing fatigue.
- Hospitalization data (particularly from private sector).
- Linking hospitalized cases with INSACOG labs for the assessment of variant-associated disease severity.

(a) Laboratory:
  (i) Genome sequencing.
  (ii) Airport surveillance, genome sequencing, and community impact.
  (iii) Development and implementation of rapid onsite test in containment zones with high transmission rates and infection load.
  (iv) Logistics for transportation of samples from field to labs.
  (v) Sewage/wastewater surveillance and creation of network in around 50 cities for proactive identification of areas potential for high transmission.

(b) IT tools:
  (i) Coordination of ministries for the development of new tools like Aarogya setu and Itihas for assisting surveillance and monitoring clusters.

(c) Nonpharmaceutical interventions:
  (i) Closure of schools and establishments.
  (ii) Enforcement of CAB and monitoring of containment zones.

1. Post-Delta wave scenario:
   (a) Public health needs:
     (i) AAR for learnings.
     (ii) Incorporation of new laboratory techniques and epidemiological tools.
     (iii) Revisiting epidemiological tools: use of IT, WOG, and WOS approach, One Health research, PHEOC, etc.
   (b) Data uploading and surveillance in many states with decreasing transmission.
   (c) Linking of community surveillance and hospitalization data.
   (d) Monitoring the new emerging variants by INSACOG labs, particularly relevant from genomic and public health perspectives.
   (e) Long covid and surveillance.

Table 4.1 Key challenges faced and measures taken

Initial phase	Public health strategies	Practical measures	Additional linked measures
	Surveillance and testing at points of entry	Arriving international travels: Screening through thermal scanners and travel details	Quarantine at POE and follow-up in the community during the incubation period
	Quarantine of suspects and isolation of case practices for ensuring the control of transmission	Appropriate health facilities for quarantine and isolation with infection control practices by use of PPEs, disinfection, and sanitization	Referred with POE-linked designated hospitals
Data analytics and use of epidemiological intelligence for reviewing transmission dynamics, period of infectiousness, generating trends and trajectories, and advance preparedness in at-risk areas.			
	Laboratory aspects	Issues and challenges	
	Development of testing facilities	Availability of primers and probes for testing kits. Availability of VTM, logistics for transportation of samples from field to labs Trained HR	
	Coordination among various ministries and media	IEC, management of infodemic and risk communication Community engagement with the availability of latest scientific knowledge	Call centers, press briefings, use of electronic and print media and pamphlets, and education material including advisories and guidelines
	IEC	Covid-appropriate behavior: Use of face mask, sanitizers, and social and physical distancing	To mitigate community transmission
	IT tools	Development of new S3 portal	Real-time surveillance data from village availability for planning interventions and strategies
Mid-phase	Public health	Challenges and strategies	
	Sero-surveys in exposed communities	Monitoring the level of antibodies in infected	For (in mild, moderate, and severe illness), exposure and assess the protective immunity

4.2 Conclusions

COVID 19 pandemic has been one of the most complex in the last many decades, which has challenged nearly all aspects of health system community, public health surveillance system (IDSP), laboratory-diagnostic competencies and capacities, quarantine and isolation facilities, healthcare workers including clinicians, non-availability of hospital including beds with isolation and infection control facilities, appropriate drugs for treatment and necessary logistics, and support therapies including availability of oxygen. The course of pandemic was marked by periods of surge driven by new variants with high community transmission, increased severity, and high mortality. The assessment of course and prediction of surges was totally uncertain. Antibody (due to exposure, vaccine, etc.) levels which were thought to be enough for the development of herd immunity levels, including various scientific modeling, provided little help to the scientific community.

India's integrated disease surveillance network, which has dedicated surveillance workforce up to district and village levels, has been one of the key strengths in initial months. Country-wide lockdown in initial days was a great support to develop self-sufficiency in PPEs, face masks, and diagnostic lab network. Aggressive whole-of-government approach driven by all ministries under the leadership of Honorable PM, revamping of IDSP surveillance portal for surveillance data, use of new technologies like genome sequencing by the establishment of INSACOG to complement community interventions, and development of effective vaccines in a short time helped the country put up a remarkable response to the pandemic.

Competing Interest The authors have no conflicts of interest to declare that are relevant to the content of this chapter.

The Diagnostic Success Story: C-CAMP COVID-19 Innovations Deployment Accelerator (C-CIDA)

5

Together India Shall

Taslimarif Saiyed, Nandita Chandavarkar, Debarshini Chakraborty, Neelanjana Janardan, Mohammad Atif Alam, Paresh Patel, Preksha Gupta, Vishal Bhardwaj, Bhavisha Wala, Niranjan Joshi, Titash Mukherjee, Kiran Vuppala, Amritanjali Kiran, Deboleena Raha Das, Neeraj Jain, Manoj Kumar, Subhashini Chandran, Madhusudhana Rao, Madhav Joshi, Abhishek Kakkar, and Sekhar Khirani

5.1 Introduction

Novel coronavirus COVID-19 emerged as one of the gravest public healthcare challenges in recent history. In March 2020, it was clear that India with its high population density would be especially susceptible. India has traditionally relied very heavily on imports of healthcare solutions like medical devices, diagnostics, technologies, etc. The unprecedented global disruption caused by the pandemic, however, directly affected these supply chains, and we were faced with a phenomenally unprecedented task of tackling a challenge of massive proportions with a glaring paucity of resources. This was the time to look within our own backyard to find deployment-ready innovations and give them the last push or nudge to reach the market.

Even before the first lockdown started, we at C-CAMP had already felt the need to identify and work with such indigenously developed and indigenously produced COVID-19 innovations for rapid deployment. One of the first such innovations we

T. Saiyed (✉) · N. Chandavarkar · D. Chakraborty · N. Janardan · M. A. Alam · P. Patel
P. Gupta · V. Bhardwaj · B. Wala · N. Joshi · T. Mukherjee · K. Vuppala · A. Kiran · D. R. Das
N. Jain · M. Kumar · S. Chandran · M. Rao · M. Joshi · A. Kakkar · S. Khirani
Centre for Cellular and Molecular Platforms (C-CAMP), Bengaluru, Karnataka, India
e-mail: taslim@ccamp.res.in

© The Editor(s) (if applicable) and The Author(s), under exclusive license to Springer Nature Singapore Pte Ltd. 2025

R. Swarup et al. (eds.), *At War with the Single Strand*,
https://doi.org/10.1007/978-981-96-8122-8_5

fast-tracked in early March was a rapid diagnostics kit that has since received regulatory nods and was deployed in the field.

As the virus marched on with no sign of reprieve, we recognized the importance of being more organized and focused in our efforts. Through the last decade or so of India's biotech boom, India's bio start-ups have been putting forward Made-in-India globally competitive solutions. Now was the time to capitalize on these technologies and align them with unmet needs for rapid deployment at an unprecedented scale.

C-CIDA or the C-CAMP COVID-19 Innovations Deployment Accelerator was born of this push in late March 2020, barely a day into lockdown with the aim to identify near-deployment-ready innovations and help them overcome last-mile issues wherever and howsoever possible.

5.2 Objective

Given the unprecedented nature of the problem, C-CIDA focused on the *innovations that were ready to be deployed immediately or in the near future* to curb the COVID-19 pandemic. Based on the assessment, C-CIDA helped in fast-tracking promising innovations by assisting in the following:

1. Pilot deployment, operational validation, and field implementation.
2. Scaling through industry and manufacturing partnerships.
3. Navigating regulatory pathway.
4. Connecting with investors and fundraising.

5.3 C-CIDA Framework

C-CIDA's nationwide call for applications was launched on March 25, 2020. As illustrated in Fig. 5.1, the whole framework of C-CIDA was designed to be extremely time-sensitive in order to have short turnaround times for identifying and fast-tracking lifesaving innovations. Applications received by C-CIDA were screened daily to prepare the first shortlist. Despite the time sensitivity, stringent assessment criteria were used to identify the innovations: *readiness for deployment, scientific*

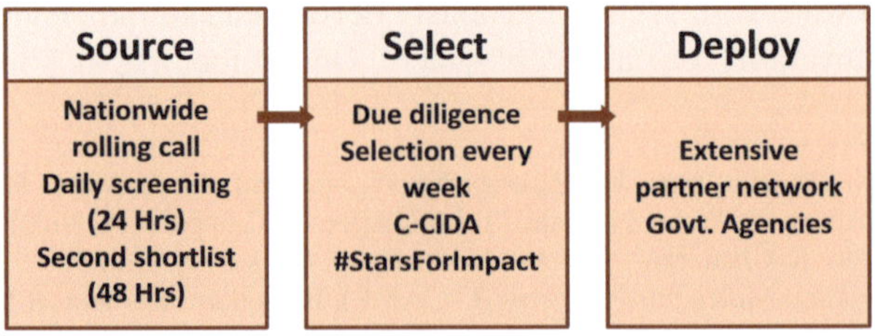

Fig. 5.1 C-CIDA's approach for a national outreach-based sourcing, due-diligence-based selection and work with government agencies for deployment of the most suitable Covid-19 solutions

validity, feasibility, and impact. Any additional information requested by reviewers was obtained from the applicants over the next day, and the second shortlist was prepared within 48 h from the receipt of applications. Over the next 2– days, due diligence of the second shortlist applications was undertaken, and the most promising innovations were selected every week.

A month down the line, C-CIDA had already received 1100+ applications, out of which *30 innovations were selected in critical categories such as diagnostic kits, ventilators, remote monitoring, preventive technologies, and more. We called them "Stars For Impact" in light of their huge potential for impact on India's COVID-19 combat.* The selected start-ups were provided assistance in aforementioned areas.

The program attracted attention not only from Indian innovation community but also from several Indian and international investor bodies and innovation centers. We joined hands with *Social Alpha, United Nations Health Innovation Exchange or UNHIE, MedTechConnect, Xynteo India2022, AIC-CCMB, CCMB, India Health Fund, PATH, Action Covid-19 Team (ACT), and 91 Springboard*. It also became a platform for Indian innovators to rise to the Honorable PM's call for a more self-reliant India or *Aatma Nirbhar Bharat*.

5.4 C-CIDA Phase 2: Scale-Up of Lifesaving Technologies

The C-CIDA initiative was started with the singular objective of combating COVID-19 pandemic through deployable deep-tech innovations. As India continued to grapple with a vicious Second Wave, C-CIDA shifted focus from harvesting innovations to rapidly scaling up the deployment of lifesaving technologies across the country. Nine such innovation deployment projects supported by diverse stakeholders were undertaken. The innovations included CPAP devices for noninvasive ventilation, contactless vital parameter monitors, last-mile cold chain devices for vaccines, oxygen concentrators, and increasing hospital bed capacity across five states through model COVID-19 care units. Another project in the training/public awareness domain was also undertaken for the development of training manuals for COVID-19 vaccination in regional languages for scaling India's vaccination program.

During the pandemic period itself, 22 C-CIDA innovations reached the market and touched an estimated 1 million+ lives in the COVID-19 fightback.

5.5 Future Directions

Through the C-CIDA platform, C-CAMP and the larger start-up ecosystem have demonstrated that where there is a will, there is a way, especially through concerted collaborative effort. Indeed, we believed in—"*Together India shall!*", and we worked toward it.

Crisis is often looked at as a fertile ground for innovation thinking. The country's COVID-19 pandemic situation has brought forth the resilience, agility, and robustness of India's innovation and start-up ecosystem. And behind the start-ups, it is the vision of innovation catalysts like C-CAMP that is helping sharpen *India's efforts of making quality health care accessible and affordable at scale!*

Competing Interest The authors have no conflicts of interest to declare that are relevant to the content of this chapter.

The Diagnostic Success Story: Pandemic Pioneers: The AMTZ Story

6

Jitendra Sharma and Mohini Mehta

Over the past five years, the medical device market in India has seen significant developments, with both imports and exports playing a crucial role in shaping its growth. The value of medical devices brought into India steadily increased from INR 40,690.00 crores in 2018–19 to INR 60,554.70 crores in 2022–23, translating to a compound annual growth rate (CAGR) of 10.45%. However, in the most recent years, there was a steady decline of −3.95% compared to the previous year, indicating the slowdown in the import sector.

On the other hand, the export market for Indian medical devices experienced robust growth. The value of medical device exports surged from INR 16,889.09 crores in 2018–19 to INR 26,602.86 crores in 2022–23, reflecting a CAGR of 12.03%. The standout year was 2022–23, which saw a remarkable year-on-year (YoY) growth of 14.20%, signaling a surge in international demand for Indian medical devices [1].

When considering the overall medical device market in India, which combines imports and exports, it becomes evident that the sector is rapidly expanding. The total market value rose from INR 57,579.09 crores in 2018–19 to INR 87,157.56 crores in 2022–23, indicating an impressive CAGR of 10.92%, while the YoY growth in 2022–23 was relatively modest at 0.95%, signifying a consistent upward trajectory for the market (Table 6.1).

These numbers paint a picture of a dynamic and evolving medical device market in India. The import sector shows signs of a steady slowdown (Fig. 6.1), while the export market demonstrates resilience and strong growth, indicating a rising global demand for Indian medical devices. The overall imports and exports of medical devices for the last five financial years are given in Fig. 6.2.

J. Sharma (✉) · M. Mehta
AMTZ Campus, Near Steel Plant, Visakhapatnam, Andhra Pradesh, India
e-mail: managingdirector@amtz.in

© The Editor(s) (if applicable) and The Author(s), under exclusive license to Springer Nature Singapore Pte Ltd. 2025

R. Swarup et al. (eds.), *At War with the Single Strand*,
https://doi.org/10.1007/978-981-96-8122-8_6

Table 6.1 YoY data of medical devices' import and export in India

Medical device market	2018–19 (INR Cr)	2019–20 (INR Cr)	2020–21 (INR Cr)	2021–22 (INR Cr)	2022–23 (INR Cr)	YoY (2022–23)	CAGR (%)
Indian medical device import market	40,690.00	41,600.84	44,580.79	63,042.57	60,554.70	−3.95%	10.45%
Indian medical device export market	16,889.09	18,204.30	19,476.49	23,295.20	26,602.86	14.20%	12.03%
Total	57,579.09	59,805.15	64,057.28	86,337.77	87,157.56	0.95%	10.92%

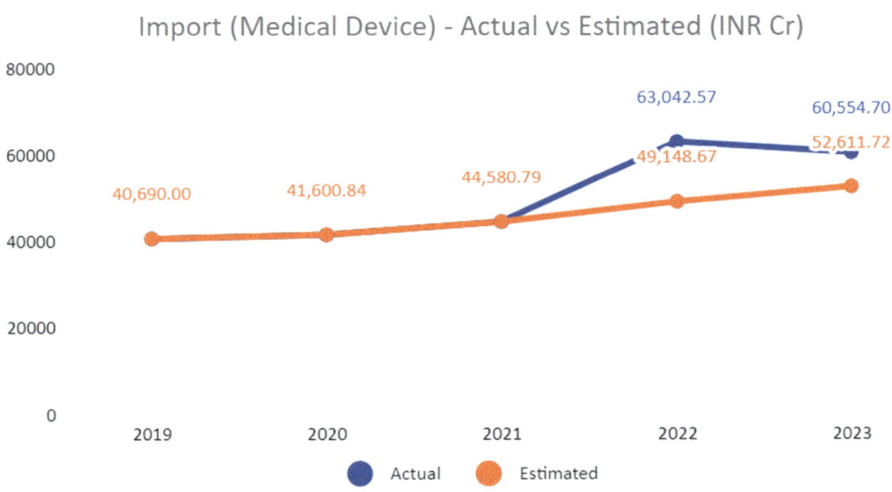

Fig. 6.1 Actual vs estimated medical device import in India (2018–2023)

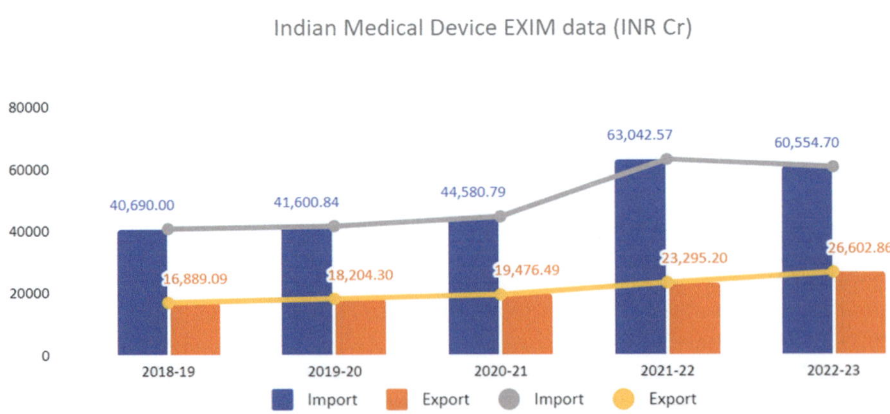

Fig. 6.2 EXIM data of the medical device industry in India (2018–2023)

As the sector continues to grow, it presents opportunities for both domestic and international players to contribute to India's healthcare landscape and tap into its thriving medical device market, and AMTZ has been a key focal in this journey of growth.

In March 2020, when India was at the brink of the COVID-19 pandemic, a pivotal moment unfolded as Prof. VijayRaghavan, the principal scientific advisor to the Government of India, convened a critical meeting with a group of experts from various scientific institutions from across the country. The purpose of this was to share ongoing plans and strategize the rapid production of scientific medical supplies. Recognizing the need to act swiftly, Prof. VijayRaghavan called upon the forum to draft a proposal for scaling up the manufacturing of ventilators and RT-PCR kits.

At that time, with the exception of Italy, which had already experienced a significant increase in COVID-19 cases and fatalities, most countries had not yet witnessed the full force of the pandemic. However, Prof. VijayRaghavan sensed the gravity of the emerging situation. With conviction and persuasion, he rallied the participants, ensuring that they comprehended the rapidly changing circumstances and were prepared to respond effectively, leveraging their scientific strengths to address the forthcoming challenges posed by COVID-19.

As an organization dedicated to advancing medical technology and manufacturing in India, the Andhra Pradesh MedTech Zone (AMTZ) under the able leadership of the MD and CEO, Dr. Jitendra Sharma, recognized the gravity of the situation and swiftly identified the immediate priorities.

Ventilators, personal protective equipment (PPE), N95 masks, infrared nontouch thermometers, Bi-PAP and C-PAP devices, multiparameter physiological monitors, and RT-PCR kits for expanding testing capabilities were among the crucial medical products that needed to be urgently produced and made available in large quantities. The existing situation in India was alarming; the country had no domestic production of ventilators, and a meager production capacity of less than 10,000 RT-PCR kits per day, exclusively for tuberculosis testing. Additionally, the daily production of PPE kits was less than 25,000, which was grossly inadequate given the projected demand.

The challenge was daunting, but AMTZ was determined to rise to the occasion and recognized that scaling up the production of these essential medical products was paramount to meet the growing demands of the healthcare system. The goal was to bridge the wide gap between the existing production capacity and the estimated demand.

Recognizing the need for swift action, Dr. Jitendra Sharma formulated the COMMAND strategy, which stood for COVID-19 MedTech Manufacturing and Development. This comprehensive plan was quickly shared with the Principal Scientific Advisor (PSA) and subsequently approved. It was then forwarded to the Department of Biotechnology (DBT) for further coordination.

Within the DBT, Dr. Renu Swarup, the secretary, had already been tirelessly working on medical technology, vaccine research, access to biological samples, and programmatic components of the COVID-19 response. Supported by the unwavering dedication of Dr. Renu and Prof. VijayRaghavan, the COMMAND proposal received preliminary approval. The plan encompassed a monumental scale-up of production machinery to facilitate the mass production of ventilators and RT-PCR diagnostic kits within a remarkably short time frame. It also involved the establishment of a nationwide supply chain and the promotion of innovation throughout the COVID-19 response.

With the initial steps in place, the stage was set for the MedTech sector in India to rise to the occasion, surmount challenges, and push boundaries. Guided by Prof. VijayRaghavan and Dr. Renu's unwavering commitment and Dr. Jitendra Sharma's visionary leadership, a collective effort of dedicated individuals took shape. This extraordinary endeavor shaped the nation's response to the COVID-19 pandemic, showcasing the resilience and determination of the Indian scientific community.

In the subsequent days, a chance conversation with Dr. P.D. Vaghela, the then Secretary of the Department of Pharmaceuticals, revealed the Ministry of Finance's collaboration with the World Bank to expedite investments in health product manufacturing. Intrigued, Dr. Sharma expressed interest in connecting with the relevant authority and was promptly introduced to the Joint Secretary in the Ministry of Finance.

The following day, while visiting the office of the Joint Secretary for guidance, Dr. Sharma was asked to wait for an hour due to an urgent meeting involving three unfamiliar gentlemen in suits. Unbeknown to them, this unexpected delay led to an extraordinary encounter. When finally invited into the office, Dr. Sharma was introduced to the two gentlemen, who happened to be the heads of the World Bank and the Asian Development Bank (ADB) in India. It was an astonishing surprise that left the room in awe.

The ADB head, familiar with the Andhra Pradesh MedTech Zone (AMTZ) project as they had financed its second phase infrastructure development, expressed tremendous appreciation for its progress. Dr. Sharma took the opportunity to explain how investing in production machinery would not only ensure product supply but also support the manufacturing of component value chains and sustainable livelihoods, particularly in the emerging field of medical technology. In a matter of minutes, the gathering wholeheartedly endorsed Dr. Sharma's request and plan. Mr. Junaid Ahmed, the Country Head of the World Bank, remarked that their organization had previously co-financed projects with the Department of Biotechnology (DBT), making this proposal an ideal fit within their guidelines. Recognizing the urgency of the unfolding global crisis, the three influential figures urged Dr. Sharma to implement this plan with utmost speed, aiming to set an example for the world.

With the World Bank onboard, the COMMAND strategy proposal swiftly reached Dr. Renu Swarup, Secretary of the DBT, who had already received the proposal from Prof. Vijay Raghavan's office. Dr. Renu Swarup wasted no time and set the wheels of assessment and action in motion, facilitating rapid implementation.

With financing secured for the scale-up machinery, a request for proposals was issued to select industry partners and innovators who would collaborate with AMTZ to accelerate development. Leveraging the state-of-the-art infrastructure and testing facilities provided by AMTZ, manufacturing was set to increase by nearly 1000 times. This remarkable endeavor gave rise to one of the largest manufacturing consortiums ever witnessed, bringing together technological giants, aspiring entrepreneurs, scientific minds, and a host of other facilitators. From the initial planning in late February to the selection of industry partners by mid-March, and project initiation on the very first day of the national lockdown (March 24, 2020), the production plan rapidly fell into place.

Within 2 weeks of the lockdown, on April 7, 2020, World Health Day, the first RT-PCR kit for COVID-19 was successfully produced at the AMTZ campus. This milestone served as a powerful testament to the unwavering commitment, strategic planning, and immense opportunities present in India's drive to establish high-scale production capabilities. It exemplified the remarkable synergy between supportive governance and progressive scientific approaches in addressing both immediate and long-term priorities.

The COMMAND Strategy adopted a three-pronged approach:

Support for Start-Ups and Innovators The DBT/BIRAC extended support to start-ups and innovators working on medical technology projects that could contribute to the COVID-19 response. These visionary individuals and entrepreneurs received assistance from AMTZ in terms of technological guidance, product realization, testing, validation, and access to prototyping and batch production facilities.

Rapid Scaling of Manufacturing The strategy recognized the potential of existing medical device manufacturers to produce critical equipment like ventilators, diagnostic kits, thermal scanners, and medical textiles, which were crucial during and after the COVID-19 period. However, scaling up manufacturing at such an accelerated pace required substantial investments in plants and machinery. The DBT-AMTZ COMMAND strategy provided a comprehensive solution by facilitating the rapid expansion of infrastructure and production capabilities.

Establishment of Standards and Safety Protocols Ensuring the quality, safety, and adherence to appropriate standards for medical technologies such as ventilators and N95 masks was of paramount importance. The strategy focused on developing validation protocols, drafting necessary standards, and facilitating their registration on the government e-marketplace (GeM). Collaborative efforts with institutions like the Bureau of Indian Standards (BIS), the Indian Council of Medical Research (ICMR), and GeM were instrumental in achieving this objective.

By addressing these three key areas, the COMMAND Strategy fostered an ecosystem that nurtured innovation, facilitated rapid manufacturing scale-up, and upheld rigorous quality and safety standards. It showcased the immense potential that arises when government agencies, scientific institutions, and industry stakeholders join forces to tackle pressing challenges and drive positive change.

On April 7, AMTZ launched the first RT-PCR kit for COVID-19, subsequently receiving the ICMR's approval of the TrueNat test. Initially producing 3000 kits in April 2020 at one production center, the AMTZ quickly ramped up its capabilities. By April 2021, across seven production centers, nearly a million kits were being manufactured per day. The AMTZ's production capacity also expanded for other essential medical supplies: from 10 ventilators a week to about 200 a day, and from 500 N95 masks a day to 100,000 N95 masks a day. It gradually became a source for a wide range of products, including infrared thermometers, multiparameter monitors, defibrillators, and even mobile container-based COVID-19 isolation ICU units and mobile RT-PCR vans.

During the national lockdown, the AMTZ campus was designated as an essential services zone. Thousands of factory workers and staff overcame numerous barriers to commute to the campus, displaying their AMTZ badges and essential service letters on their bikes or cars. The dedication and sacrifice of the staff, some residing in "hot spot" areas, who isolated themselves from their families to minimize the risk of transmission, played a crucial role. The AMTZ's environmentally friendly campus, adorned with over 200,000 plants and powered by a renewable energy solar plant, contributed to maintaining a clean and healthy environment.

Workdays at the AMTZ began at 9:00 a.m. and often continued until midnight. Overcoming logistical challenges, such as sourcing raw materials and components from various cities and countries, and addressing positive COVID-19 cases among the manufacturing lines, AMTZ managed to transport over 3000 trucks of essential medical supplies across the country (Fig. 6.3).

The fragmented logistics and supply chain globally posed significant challenges in obtaining spare parts, critical components, and raw materials for the AMTZ. One of the major time-critical bottlenecks was the importation of oligonucleotides (primers and probes) from major suppliers in the United States, such as ThermoFisher, and Europe, like Eurofins. These oligonucleotides were the key component of all RT-PCR kits. However, the AMTZ was determined to overcome this dependency and achieve self-reliance.

Fig. 6.3 Truckloads of life-saving products being shipped from AMTZ

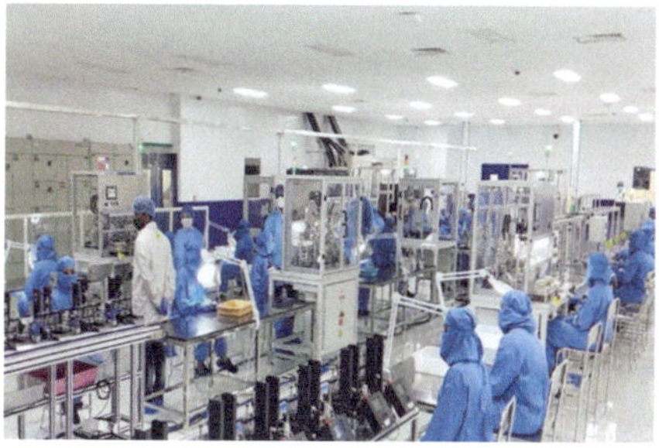

Fig. 6.4 Over 100 teams working in two to three shifts a day, displaying a strong sense of teamwork and support

The AMTZ successfully indigenized the production of oligonucleotides and began synthesizing them in-house. This breakthrough not only resolved the bottleneck but also positioned the AMTZ as a supplier of high-quality oligonucleotides. Additionally, the AMTZ expanded its capabilities by manufacturing other critical components like printed circuit boards and oxygen and flow sensors for medical electronic equipment within its own campus.

Multiple teams were formed to handle different aspects of the initiative. One team focused on mapping specifications and gathering market intelligence for each product segment, while another provided engineering design support. A third team facilitated the installation and commissioning of production machinery. HR teams worked tirelessly to recruit and train hundreds of technicians and engineers, while logistics and regulatory support teams were added to assist manufacturing partners. A comprehensive network was established to overcome logistical challenges and ensure smooth operations.

The collaboration among manufacturing partners was remarkable. Over 100 teams worked in two to three shifts a day, displaying a strong sense of teamwork and support (Fig. 6.4), Instead of competing, manufacturers embraced a collaborative approach, assisting each other with R&D and supply chain challenges. This collective effort proved vital in product development and clinical validations. Leaders from different organizations, including Satyaki Banerjee, Maninder Singh, Sashi Kumar, Rajesh Patel, Chandrasekhar Nair, Sriram, Manoj Chugh, and Manjiv Singh, worked collaboratively to ensure success and save countless lives by rapidly supplying critical medical devices.

6.1 World's Largest IVD Cluster at AMTZ—A Million RT PCR Kits a Day

Additional support was received when Mr. Ravi Capoor, the Secretary of the Ministry of Textiles, Government of India, was called upon to lead the manufacturing of medical textiles in March 2020. Within a remarkably short span of 22 days, a state-of-the-art PPE and mask making center named TEXTURA was established. The decision to initiate manufacturing was followed by the allocation of adequate resources and the efficient management of time to ensure timely deliverables for each product. These extraordinary efforts at the AMTZ set a momentum for achieving even greater milestones, fueling hope and saving lives across the nation (Fig. 6.5).

Under normal circumstances, product development follows a systematic and controlled process that typically spans several years. However, given the urgency of the pandemic, there was a need to expedite technical specifications and product engineering in a matter of weeks. A common procurement plan was devised for all manufacturing partners, and centralized procurement of critical components was implemented to enhance their collective bargaining power. This strategic approach significantly reduced delivery times. Whenever required, Indian missions abroad were approached to assist in sourcing essential components. Indian embassies and consulates in the United States, Japan, China, and various European countries provided essential coordination and support (Fig. 6.6).

Fig. 6.5 AMTZ—the largest MedTech cluster globally—that makes from mask to MRI

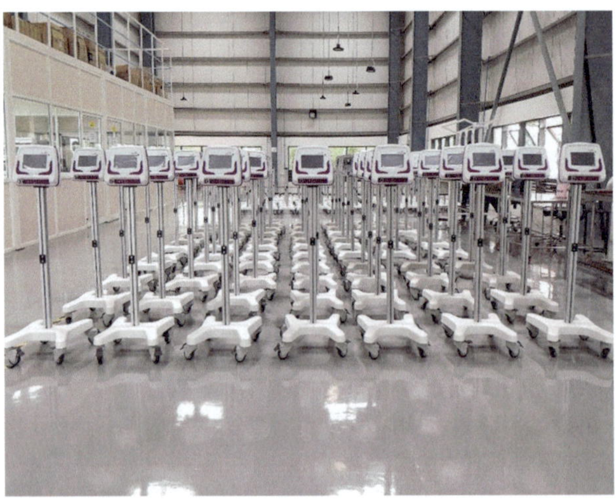

Fig. 6.6 Tens of thousands of ventilators shipped across the country

Throughout this collaborative endeavor, Dr. Arabinda Mitra, the Scientific Secretary in the Office of the Principal Scientific Advisor to the Government of India, consistently offered valuable support. Notably, Ms. Mona Khandhar from the Indian Embassy in Tokyo, Japan, and Mr. Dhananjay Tiwari at the Indian Mission in Washington DC, USA, provided tremendous assistance in facilitating the procurement process.

During the second wave of the COVID-19 pandemic, the Andhra Pradesh MedTech Zone (AMTZ) faced a significant challenge. The need for oxygen concentrators became apparent, prompting a proposal from the Empowered Group for Procurement to allocate at least 25 percent of the ventilator count for their procurement. Oxygen concentrators were deemed crucial in providing rapid support to patients, but due to the initial focus on ventilators during the first wave, their development and procurement were not prioritized.

When the second wave struck between April and August 2021, the demand for oxygen escalated to unprecedented levels. Unfortunately, the crucial component required for manufacturing oxygen concentrators, the lithium sieve or zeolite, was not domestically produced in India. This scarcity arose suddenly as major suppliers, including France, the US, Japan, and China, experienced shortages. Additionally, the urgent need for oil-free compressors further complicated the situation. Despite the AMTZ's best efforts, only a limited number of oxygen concentrators could be supplied within the first 30 days.

However, the AMTZ swiftly undertook large-scale indigenization initiatives to overcome these challenges and gradually achieved self-sufficiency by manufacturing critical components such as oil-free compressors and oxygen sensors on-site. This strategic indigenization plan allowed the AMTZ to become a significant supplier of oxygen concentrators to public sector organizations, state governments, the Indian Navy, and the Infosys Foundation. This timely supply of oxygen concentrators played a vital role in saving numerous lives (Fig. 6.7).

Fig. 6.7 Over 20,000 oxygen concentrators supplied

The eureka moment of the Indian MedTech sector emerged as medical equipment were "made in lockdown" in the global landscape [2]. This development not only showcased progress but also instilled confidence in the enterprising attitude of those dedicated to researching and manufacturing medical equipment in India. The AMTZ, as a well-established medical technology manufacturing ecosystem, possessed all the necessary elements to rise to the occasion.

The existing infrastructure within the AMTZ campus, including common scientific facilities and manufacturing resources, played a crucial role in this endeavor. Specialized laboratories, warehousing facilities, and testing centers provided a significant advantage for the mission at hand. Collaborative efforts among various entities, such as the Centre for Electromagnetic Compatibility and Safety Testing, Centre for Biomaterial Testing, Centre for 3D printing, Centre for Lasers, Gamma Irradiation, Moulding, and other industrial service centers, brought together researchers, entrepreneurs, and manufacturers under one roof (Fig. 6.8).

The convergence of these scientific and industrial entities fostered a collaborative environment where ideas could flourish, technologies could be tested, and innovations could take shape. Researchers, entrepreneurs, and manufacturers worked hand-in-hand, leveraging the shared resources and expertise available within the AMTZ ecosystem. This collaborative approach further accelerated the development and production of high-quality medical equipment.

The AMTZ's commitment to innovation, combined with the availability of common facilities, created a fertile ground for groundbreaking advancements in the MedTech sector. It paved the way for transformative discoveries and breakthroughs that contributed to the nation's healthcare infrastructure. By harnessing the potential of their collective capabilities, the AMTZ and its partners propelled India's MedTech sector forward, making significant strides in medical equipment production and research.

As the need for rapid product development grew, the demand for more systematic designs and extensive infrastructural facilities became both necessary

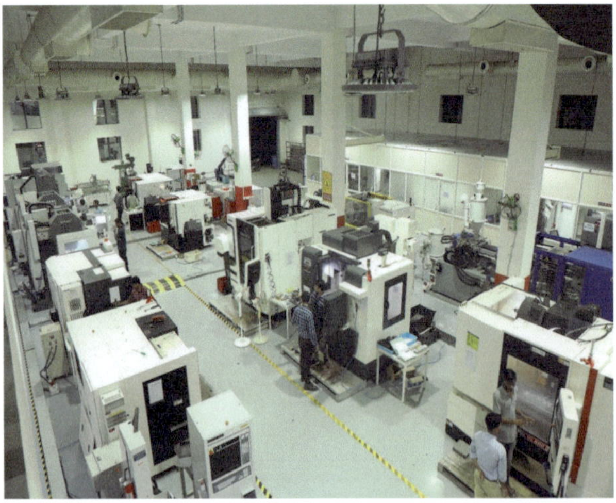

Fig. 6.8 Common scientific facilities for industry—AMTZ is a demonstration of this practice

and achievable. One such remarkable development was the introduction of the mobile diagnostic unit known as the Infectious Disease Diagnostic Lab (I-Lab), equipped with a biosafety facility capable of conducting various tests including RT-PCR, enzyme-linked immunoassay (ELISA), and more. This innovation aimed to ensure convenient testing access in rural areas where healthcare facilities were limited.

On June 18, 2020, the union health minister inaugurated the first I-Lab in New Delhi, marking a significant milestone in expanding diagnostic capabilities. The I-Lab, a Rapid Response Mobile Laboratory, stood apart due to its unique ability to perform both molecular biology and serology antibody diagnostic testing, primarily focused on infectious diseases. This fully equipped mobile unit could be deployed to different locations, enabling efficient testing for infectious diseases such as COVID-19, Hepatitis B&C, HIV, Dengue, Chikungunya, Malaria, as well as routine biochemical tests including glucose (diabetic), GTT, liver, kidney, thyroid function tests, and more.

The objectives of the I-Lab were multifold. First, it aimed to detect, monitor, and control emergency pandemic events and outbreaks promptly. Second, the mobile laboratory facilitated on-site testing and analysis, eliminating the need for sample transportation to distant testing facilities. This increased the speed and capacity for laboratory testing, ensuring faster and more efficient diagnosis. Moreover, adhering to international standards, the I-Lab incorporated diagnostic testing processes of the highest quality.

Dr. Renu Swarup, known for entrusting complex tasks to institutions she believed in, had requested the development of the mobile testing facility, and the first mobile laboratory was constructed at the AMTZ in a record time of 8 days. Since then, several states, particularly those in the challenging terrains of Northeast India, have deployed these mobile testing laboratories, serving as a valuable asset in resource-constrained settings.

The introduction of the I-Lab brought advanced diagnostic capabilities to remote areas, allowing for timely detection and containment of infectious diseases. This mobile testing facility played a crucial role in expanding access to diagnostics, particularly in underserved regions. It exemplifies the commitment of healthcare professionals and institutions like the AMTZ to deliver innovative solutions that address the unique challenges faced in healthcare delivery across the nation. The I-Lab stands as a testament to the power of swift action, collaboration, and ingenuity in the face of adversity, enabling more effective disease management and safeguarding the health and well-being of communities.

Alongside the successful establishment of the Infectious Disease Diagnostic Lab (I-Lab), the AMTZ proudly housed the ICMR Kit Validation Laboratory, a critical component of their comprehensive testing infrastructure.

Within the ICMR Kit Validation Laboratory, a dedicated team of experts took on the crucial task of testing SARS-CoV samples using rapid antigen kits and RT PCR methods. With utmost precision and adherence to guidelines laid out by the Indian Council of Medical Research (ICMR), the laboratory collected patient samples and conducted rigorous testing procedures.

To maintain the highest standards of quality and accuracy, the laboratory diligently followed the NABL standards of ISO 15189. This ensured that every step of the testing process, from sample collection to analysis, was conducted with meticulous attention to detail, instilling confidence in the reliability of the results.

Recognizing the significance of ongoing quality control, the laboratory actively participated in the External Quality Assurance Scheme (EQAS) in collaboration with an ICMR laboratory. Through regular proficiency testing and benchmarking, the laboratory continuously monitored and improved its testing processes, contributing to the overall reliability and credibility of the AMTZ's diagnostic capabilities.

Furthermore, the ICMR Kit Validation Laboratory showcased its expertise by conducting thorough validation of rapid antigen kits. Through comprehensive assessments and evaluations, the laboratory confirmed the effectiveness and accuracy of these crucial testing tools. This validation process served as an essential element of quality control, further strengthening the AMTZ's commitment to providing reliable and accurate diagnostic services.

The establishment of the ICMR Kit Validation Laboratory within the AMTZ was a testament to the institution's dedication to excellence and its unwavering commitment to the health and well-being of the people. The collaboration between skilled professionals and cutting-edge technology in these testing facilities played a vital role in combating the pandemic and ensuring the efficient and accurate diagnosis of COVID-19 cases.

In addition to this, the AMTZ embarked on a groundbreaking initiative to address the healthcare needs of remote and underserved areas. Container hospitals were introduced, which were fully equipped with a hospital-like environment, including eight beds, oxygen pipelines, supply lines, and life-saving equipment. These container hospitals, known as Safe Zones, were housed within 40x8 feet containers that could be easily transported by trucks to any location lacking established clinical settings.

The concept of container hospitals arose from a request made by Prof. Vijay Raghavan's office, recognizing the dire need for accessible healthcare in remote areas. These Safe Zones, built at the AMTZ, proved immensely valuable and were dispatched to over 50 destinations across the country, providing vital health services to communities that lacked proper medical infrastructure, effectively bridging the healthcare gap.

Another groundbreaking innovation that emerged from the AMTZ was the Oxy-Wheel, a pressure swing adsorption plant designed for oxygen generation. This plant had the capacity to meet the oxygen requirements of multiple hospitals in a single day, at a nominal cost. By eliminating the need for permanent oxygen generation plants and the associated capital investments, the Oxy-Wheel played a pivotal role in making oxygen supply more accessible and affordable for healthcare facilities.

The impact of the AMTZ's endeavors extended beyond the boundaries of the country. While catering to the domestic demand for medical equipment, AMTZ aimed to reduce India's import dependency significantly. With a target to produce medical equipment worth over 20,000 crore rupees ($2 billion) annually, the AMTZ aimed to address almost 50 percent of India's import requirements. This ambitious goal demonstrated their commitment to liberating the nation from import reliance and establishing India as a global leader in medical technology development and production.

The products manufactured at the AMTZ, such as RT-PCR kits and oxygen concentrators, not only served the domestic market but also found their way to international destinations. With their commitment to affordability, accessibility, and high-quality standards, the AMTZ products gained recognition and were exported to several countries, bolstering India's reputation as a reliable supplier of medical equipment.

Today, the AMTZ stands as a testament to its unwavering dedication to the nation's health and well-being. By supplying affordable, accessible, and superior-quality medical products across multiple states and geographical regions, the AMTZ continues to safeguard the health of the people. The relentless pursuit of reducing import dependency has not only transformed the Indian medical technology sector but also highlighted its capacity to save lives and contribute to social welfare.

The success of the AMTZ during the COVID-19 pandemic exemplifies the power of collaboration and resource utilization in achieving remarkable outcomes. It serves as a testament to the unwavering dedication of individuals and organizations committed to advancing healthcare and ensuring the availability of critical medical equipment. The accomplishments of the AMTZ and its collaborative network continue to inspire and drive the Indian MedTech sector toward a future of innovation, resilience, and improved patient care.

Competing Interest The authors have no conflicts of interest to declare that are relevant to the content of this chapter.

References

1. https://tradestat.commerce.gov.in/ftpa/
2. Made in Lockdown: Dr Jitendra Sharma.

The Diagnostic Success Story: Diagnostics in the COVID-19 Pandemic: A Triumph of Science and Innovation: *Mylab's Remarkable Role in Transforming Diagnostics During the COVID-19 Pandemic*

7

Hasmukh Rawal

7.1 Introduction

The COVID-19 pandemic, arising from the novel coronavirus SARS-CoV-2, presented an unprecedented global challenge. While it brought about substantial disruptions in health, society, and the economy, it also witnessed an extraordinary achievement in the realm of diagnostics. Diagnostics emerged as a critical pillar in the pandemic response, serving as the linchpin for identifying cases, monitoring virus spread, tracking mutations, and facilitating efficient vaccine distribution.

In March 2020, India had merely 14 approved laboratories for conducting RT-PCR tests for COVID-19. Today, that number has soared to over 3000 labs, encompassing both public and private sectors. With a staggering 931 million tests conducted until June 2023, India's diagnostic journey during the pandemic stands out as one of the world's most remarkable success stories.

The diagnostic sector witnessed a profound transformation and numerous innovations during the COVID-19 pandemic. As the pandemic escalated, various diagnostic entities began domesticating the development of diagnostic solutions in India, significantly contributing to making clinical decisions easier and aiding patients during the crisis.

H. Rawal (✉)
Mylab Discovery Solutions Private Limited, Lonavala Industrial Co-Operative Estate Ltd, Nangargaon Lonawala, Pune, Maharashtra, India
e-mail: hasmukh@mylabglobal.com

© The Editor(s) (if applicable) and The Author(s), under exclusive license to Springer Nature Singapore Pte Ltd. 2025

R. Swarup et al. (eds.), *At War with the Single Strand*,
https://doi.org/10.1007/978-981-96-8122-8_7

7.2 Accelerating Scientific Innovations

COVID-19 injected an unprecedented sense of urgency into the progression of scientific innovations, hastening the journey from the inception of an idea or prototype to the large-scale production and real-world implementation of these groundbreaking innovations. It ushered in a new era of diagnostic testing, with a spectrum of screening and diagnostic tests swiftly emerging. These innovations span from blood-based rapid diagnostic tests to the more precise RT-PCR-based methods designed to detect the genetic material of the novel coronavirus. Among these diverse contributions to research, some have managed to achieve a delicate equilibrium between rapid detection and high accuracy, representing a remarkable stride forward in the field of diagnostics.

Following the second wave of the pandemic in India, diagnostics witnessed a shift from centralized laboratories to decentralized point-of-care (POC) testing. During the peak of the Delta variant wave in 2021, an average of 1.8 million tests were conducted daily, a feat made possible by local manufacturing and investments in diagnostic research and development.

The pandemic catalyzed significant advancements in miniaturized and simplified molecular diagnostic platforms. These technologies can be decentralized and deployed in primary health clinics or even homes, enhancing clinical care and surveillance systems through data-sharing capabilities.

Furthermore, the normalization of self-testing and home-based testing during the pandemic has been pivotal. Tests were administered not only in homes but also in schools and workplaces, empowering individuals to take control of their health. This newfound familiarity with self-testing, coupled with technological progress and streamlined regulatory pathways, could revolutionize healthcare, particularly for diseases where stigma hampers access, such as sexually transmitted infections, HIV, or tuberculosis. Self-testing for airborne respiratory infections, including COVID-19, flu, and RSV, may also enable more informed decisions regarding self-isolation.

However, these achievements were not without their challenges. During the initial stages of the pandemic, as emphasized by the World Health Organization (WHO), testing was of paramount importance. India grappled with a surging demand for diagnostics. Shortages were not limited to test kits alone but extended to ancillary support such as personal protective equipment (PPE) kits and viral transport media, vital for safely transporting testing samples. The country was reliant on importing kits, which not only drove up costs but also caused delays in procurement.

Despite the myriad diagnostic kits available in the market, affordability and accessibility remain significant challenges, in India. Several factors contribute to this, including the unavailability of necessary reagents, a lack of technical expertise for performing tests and operating equipment, and limited access to research facilities for procuring virus samples. Meeting medical requirements concerning the speed, volume, and cost of diagnostic tools also faced significant constraints.

Addressing the multifaceted requirements for detecting, diagnosing, and responding to threats like the pandemic necessitates the mobilization of scientific research and technical capacity on a massive scale.

Indian researchers, innovators, and regulators have admirably collaborated to introduce low-cost diagnostics, develop therapeutics, and conduct research aimed at creating safe and effective vaccines during the pandemic.

7.2.1 Mylab: An Innovative Force in Pandemic

Mylab, a key player in the Indian diagnostic landscape, emerged as an innovative force during the pandemic. When the pandemic hit India, the country lacked domestically manufactured COVID-19 RT-PCR test kits, relying entirely on imports. Mylab rose to the occasion, becoming the first Indian company to develop an indigenous RT-PCR test kit for COVID-19 in a record time of 6 weeks since the outbreak. Subsequently, Mylab launched several COVID-testing-related products, including India's first self-test kit, CoviSelf. The company's unwavering commitment to increasing India's testing capacity was evident throughout the pandemic. Mylab's impact during the pandemic was significant, with over 50 million people tested using their RT-PCR kit for COVID-19.

7.3 From Import Dependence to Indigenous Innovation: Anticipating the Need for Indigenous Solutions

In January 2020, Mylab's executives foresaw the potential arrival of the pandemic in India and recognized the imperative of developing and manufacturing COVID-19 RT-PCR tests. Mylab was well-positioned for this endeavor, having previously developed test kits for various viral infections, including HIV, HBV, and HCV.

Indeed, navigating the complexities of developing a diagnostic test for a biohazardous virus like SARS-CoV-2 presented unique challenges. While the actual virus sample remained inaccessible to companies like Mylab, they relied on a combination of valuable resources, guidance, and their pre-established capabilities to embark on this critical mission.

Although Mylab did not have access to the virus's actual sample due to its biohazardous nature, the company drew upon guidelines from the Centers for Disease Control and Prevention (CDC) and the World Health Organization (WHO) to formulate its test development strategy. Mylab simultaneously worked on three test formulations, ultimately submitting the most promising one to the government by March 2020.*

Maintaining rigorous quality control criteria was a fundamental aspect of Mylab's approach. Ensuring the accuracy and reliability of the diagnostic test was paramount. Crucially, one key advantage that Mylab possessed was its prepandemic preparations. Even before the COVID-19 outbreak, the company had established a robust manufacturing unit spanning over 30,000 square feet. Furthermore, Mylab had secured approvals from local FDA and CDSCO for manufacturing Class C and Class D diagnostic kits for critical molecular diagnostic products (HIV, HBV, HCV, NAT, and others) even before the pandemic struck. This readiness allowed Mylab to swiftly redirect its resources toward the production of COVID-19 test kits, minimizing the time it took to transition from planning to execution.

* Press release: Fast track approval for Indian COVID-19 testing kits for commercial use. https://www.icmr.gov.in/pdf/press_realease_files/ICMR_Press_Release_23032020.pdf

7.4 Innovative Solutions that Redefined Diagnostics

The urgency created by the pandemic spurred Mylab to accelerate the development of affordable, scalable, and rapid medical technologies. Mylab's "PathoDetect CoVID-19 detection kit" was a prime example of such innovation. The kit implemented a robust diagnostic methodology for use in public health laboratories, offering screening and detection of COVID-19 with unparalleled efficiency. The workflow reduced the detection time to just 2.5 h, compared to more than 7 hours required by other protocols. Moreover, the kit was introduced at a quarter of the price of competing PCR kits, underscoring Mylab's focus on social impact.

7.5 Overcoming Logistical Challenges Through Collaboration

Mylab received its commercial license for manufacturing and selling COVID-19 test kits in March 2020. Initially, Mylab relied on manual processes for kit production, with a capacity of merely 10,000 tests per week. While Mylab had the expertise and innovative drive to produce tests, it faced logistical challenges that threatened to impede progress. The countrywide lockdown, coupled with global shortages of specific reagents, posed significant challenges in procuring raw materials. Some reagents crucial for the test kits were unavailable due to their short shelf life and the practice of manufacturing them on-demand. These reagents were supplied by both local Indian manufacturers and multinational corporations, who were Mylab's competitors. As a result, Mylab had to procure certain reagents at significantly higher prices to meet the demand.

Mylab's extensive experience in distributing life science research products became an invaluable asset in navigating these challenges. The company needed to secure a steady supply of raw materials for test kit production on a large scale. Mylab's established connections with companies in the United States, Europe, and other countries facilitated the procurement of essential supplies. Collaboration with the Ministry of Health sensitized foreign manufacturers to the criticality of Mylab's requirements. The Ministry of Civil Aviation assisted in the transportation of components and materials from various global locations to Mylab's facility in Lonavala, Pune. Effective coordination between government authorities, suppliers, and Mylab's team enabled the rapid scaling of production and distribution.

Mylab forged a crucial partnership with Syngene International, instrumental in localizing the supply of raw materials. This collaboration addressed shortages and long lead times for critical components in RT-PCR kits. Syngene's efforts enabled Mylab and other RT-PCR kit manufacturers to rely on a dependable, cost-effective, and local supplier who prioritized domestic orders.

7.6 Regulatory Expediency

Obtaining regulatory approval for test kits involved a two-stage process. The first stage required approval from the Indian Council of Medical Research (ICMR), which clinically evaluated the product. The second stage involved regulatory approval from the Central Drugs Standard Control Organization (CDSCO), which assessed manufacturing standard operating procedures, quality checks, and the associated workforce.

Mylab achieved a significant milestone by becoming the first organization to secure certification from the Indian Council of Medical Research (ICMR). Even before attaining ICMR's approval, Mylab took proactive steps by inviting the Central Drugs Standard Control Organization (CDSCO) to audit their facility. This strategic move aimed to ensure that they could swiftly initiate the manufacturing of test kits as soon as ICMR granted its approval. CDSCO cooperated with this request, allowing Mylab to commence the production of test kits as soon as both approvals were in place. This collaborative approach was indicative of the regulators' recognition of the urgency of the situation.

Both product and facility approvals were expedited, and notably, they were granted on the same date. This expeditious process showcased the dedication of regulatory authorities in expediting the deployment of critical diagnostic tools. The support extended beyond the approval phase, as regulatory and other government bodies continued to provide guidance, inquiries, and assistance even after the test kit received approval.

In addition to regulatory support, Mylab received valuable assistance in logistical matters related to the importation of raw materials. India Post also played a crucial role by approaching Mylab and facilitating the distribution of COVID-19 RT-PCR test kits throughout the country. This collaborative effort between regulatory bodies and government entities demonstrated a unified commitment to addressing the challenges posed by the pandemic and ensuring the rapid availability of essential diagnostic resources to the nation.

7.7 Scaling Up Testing Capacity in India

Mylab's commitment to innovation and adaptability extended beyond test kit development. Recognizing the need for rapid, large-scale testing, the company embarked on an ambitious mission to scale up its manufacturing capacity. Within 2 months of obtaining approval, Mylab increased its testing capacity from a mere 10,000 tests per week to a staggering 1.2 million tests per week. Automation played a pivotal role in this transformation, as Mylab sought to automate its processes but encountered logistical challenges. The company purchased six robotic systems from American companies, four of which were installed within a month. The Indian government facilitated the movement of essential equipment and helped Mylab acquire critical raw materials stuck in Germany and Singapore. Consequently, Mylab's testing capacity surged from 10,000 tests to 550,000 tests per week, thanks to automation.

Anticipating future needs, Mylab developed the Rapid Antigen Test (RAT), which could be conducted outside traditional laboratory settings, such as in field health camps. RATs were easier to administer and provided relatively faster results. Mylab's foresight led the company to acquire a 16-acre land adjacent to its existing manufacturing facility to construct a larger manufacturing unit. In June 2020, Mylab obtained a commercial license for RATs and set up a new facility, bolstering India's capacity to conduct RATs significantly. The facility's capacity surged from 1.5 lakh tests per day to 15 lakh tests per day by October 2021. This strategic expansion into RAT production proved prescient during India's second COVID-19 wave in 2021, when infection rates rose steeply.

Home tests facilitated mass screening for COVID-19, delivering rapid results at a lower cost, alleviating the burden on overwhelmed laboratories, reducing delays in testing, and minimizing the strain on healthcare workers who collected samples from homes. Mylab saw the opportunity and was the first company to introduce the first self-test kit, CoviSelf in India.

Additionally, Mylab also introduced "Compact XL," India's first fully automated sample-to-PCR-ready system for molecular diagnostics. This system, capable of conducting up to 96 tests in a single run (equivalent to over 1000 tests per day), was designed to automate laboratory operations and reduce the need for manual intervention.

To address the substantial backlog in RT-PCR testing during the second wave of COVID-19 in India, Mylab deployed a fleet of ICMR-approved and NABL-certified mobile testing labs. These mobile labs, equipped with automated Compact XL machines, conducted tests three times faster than conventional labs, owing to parallel processing and automated handling. These mobile labs could process over 1500 to 3000 samples per day, serving areas with high caseloads and traveling to rural regions with limited access to testing facilities.

Recognizing the importance of rapid and efficient testing to curb the spread of the virus, Mylab collaborated with airport authorities to set up on-site testing centers. The company established state-of-the-art testing facilities at several airports.

These testing facilities, strategically located at airports across the country, provided travellers with convenient access to COVID-19 testing services. Passengers arriving at or departing from these airports could undergo RT-PCR tests, rapid antigen tests, or other authorized diagnostic procedures to ensure their safety and the safety of others during air travel.

Mylab's involvement in airport testing not only facilitated safe travel but also contributed to the broader goal of minimizing the virus's transmission. By offering quick and reliable testing services, Mylab supported the efforts to identify and isolate potential COVID-19 cases promptly, thus helping to prevent outbreaks and protect public health.

These airport testing facilities were part of Mylab's broader commitment to leveraging diagnostic innovations for the betterment of society, healthcare providers, and laboratories.

7.8 A Culture of Responsibility and Innovation

The initial hesitation among staff to work on the novel coronavirus, driven by concerns over its unknown nature, eventually gave way to a sense of responsibility. Mylab implemented rigorous screening schedules, conducting weekly RT-PCR tests for the entire team. During the initial lockdown stages, the company redirected its research and development staff to assist in manufacturing. Mylab expanded its teams to meet the surging demand during the pandemic. The company hired several interns as temporary workers, increasing the factory's workforce tenfold compared to pre-COVID times.

Mylab's commitment to innovation and its ability to adapt to evolving pandemic challenges were pivotal in its contribution to India's diagnostics success story. Figures 7.1, 7.2, 7.3, and 7.4 show the different COVID kits developed. Table 7.1 gives a timeline of the production of these kits.

Fig. 7.1 PathoDetect (COVID-19 RTPCR test)

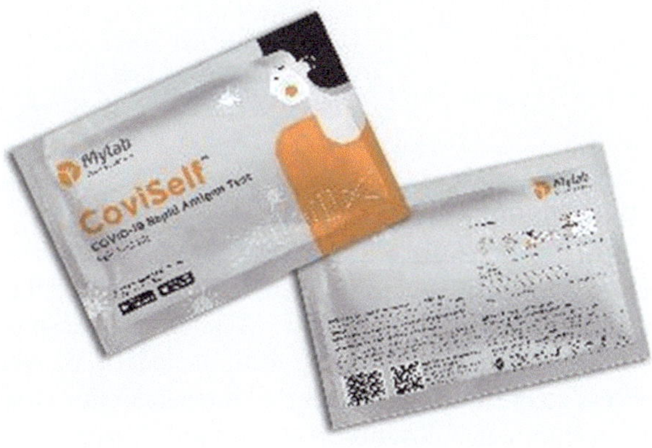

Fig. 7.2 CoviSelf (COVID-19 self-test)

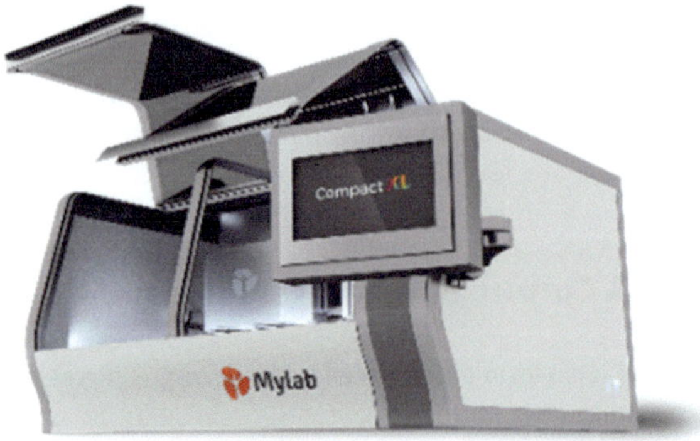

Fig. 7.3 *Compact XL*—fully automated sample-to-PCR-ready system

Fig. 7.4 CoviSwift (rapid RT-PCR, PoC testing kit)

Table 7.1 Timeline for COVID product launches

Year	COVID solutions from Mylab
2020	*PathoDetect (COVID-19 RTPCR test)* *Compact XL* *PathoCatch (rapid antigen testing kit)* *ViroTrap (molecular transport medium)*
2021	*CoviSelf (COVID-19 self -est kit)* *Mylab mobile testing vans*
2022	*CoviSwift (Rapid RT-PCR, PoC testing solution)*

7.9 Conclusions

In summary, the COVID-19 pandemic underscored the critical role of diagnostics in managing a global health crisis. Mylab's rapid response, innovative solutions, and commitment to expanding the testing capacity played a significant role in India's success story in diagnostics during the pandemic. Their efforts exemplify the power of collaboration, innovation, and adaptability in addressing unprecedented challenges.

Mylab's journey from import dependence to indigenous innovation serves as an inspiring example of how science and determination can triumph over adversity, providing hope and valuable lessons for future global health crises.

Competing Interest The authors have no conflicts of interest to declare that are relevant to the content of this chapter.

8 Make in India Vaccines: A True Story of Atmanirbhar Bharat: The Race to Pioneer a Solution to the RNA Challenge

V. Krishna Mohan, Brunda Ganneru, Siddharth Reddy Chiteti, V. K. Srinivas, and Partha Kumar Sarkar

8.1 Bharat Biotech

In the year 2020, the 'novel' virus kept us contained in our homes for months—re-establishing our relationship with the outside world, the government and even each other. This mysterious virus claimed millions of lives, weakened economies and forced businesses and borders shut. Paradoxically, it presented a revived appreciation for the outdoors, life's simple pleasure, and less polarisation—the fight was us against the virus.

To some people, the advent of a pandemic was a grim inevitability and was not shocking. These unsurprised people were scientists. Those schooled in infectious diseases declared that the virus genome was written in RNA, not DNA.

The strong commitment of the scientific community to addressing COVID-19 disease led to the publication of 27,255 manuscripts by June 2020. This is just as essential as the actual development of a vaccine because potential therapeutic approaches require understanding the virus' mechanism of action. Academic institutions, biotech companies and vaccine manufacturers rose to the challenge and started their race against time. After thorough deliberation, the most extensive portfolio of vaccines was created using several platform technologies—inactivated virus, mRNA, DNA, protein subunit and other newer technologies.

V. K. Mohan (✉) · B. Ganneru · S. R. Chiteti · V. K. Srinivas · P. K. Sarkar
Bharat Biotech, Shamirpet, Hyderabad, India
e-mail: kmohan@bharatbiotech.com

© The Editor(s) (if applicable) and The Author(s), under exclusive license to Springer Nature Singapore Pte Ltd. 2025
R. Swarup et al. (eds.), *At War with the Single Strand*,
https://doi.org/10.1007/978-981-96-8122-8_8

8.2 A Beacon of Hope: Bharat Biotech

The unprecedented reality of the COVID-19 pandemic had left the world hanging on the last thread. Bearing in mind its vision to offer healthcare solutions to humankind's most dreaded illnesses, Bharat Biotech stepped up. Bharat Biotech International Limited (BBIL), a Hyderabad-based biotechnology company with a rich history of vaccine development, emerged as a beacon of hope. Established by Dr. Krishna Ella and Ms. Suchitra Ella in 1996, Bharat Biotech has a track record of pioneering research in the field.

The journey of COVAXIN® began with Bharat Biotech Chairman Dr. Krishna Ella writing a letter to ICMR requesting the deadly SARS-COV-2 strain. The highly feared live virus was fearlessly transported overnight from ICMR-National Institute of Virology (NIV) laboratories through imminent restrictions on the backseat of a car to the Bharat Biotech campus, where it was cultivated and processed as per regulatory requirements.

The production process for COVAXIN® was rigorous and demanding. The viral strain underwent bioprocessing in a high-containment facility where it was grown, inactivated and purified to make a whole-virion-inactivated COVID-19 vaccine. The vaccine candidate was then evaluated for safety, immunogenicity and efficacy in animal models and, eventually, humans.

8.3 The Power of Vaccine Inactivation

COVAXIN®, also known as **BBV152**, was conceived as a potential solution to the global health crisis. The vaccine was developed using an inactivated virus platform. This method involves growing the SARS-COV-2 virus in a controlled environment and then inactivating it using a chemical called beta-propiolactone in a Biosafety Level 3 (BSL-3) facility, which is GMP-certified. The inactivated whole-virion vaccine platform is the world's most well-established manufacturing platform for vaccine production. Incidentally, this approach was established to create vaccines at the company, as demonstrated earlier through their highly successful Rabies and Japanese Encephalitis vaccines.

8.4 Adjuvanation: Enhancing Immunity and Boosting Protection

The efficacy of a vaccine depends not only on the antigen components but also on adjuvants. An adjuvant is an ingredient in some vaccines that stimulates a stronger immune response in individuals who receive it. In other words, they help vaccines work better. BBV152 was formulated with an imidazoquinolinone molecule (IMDG) absorbed on alum (also called Algel) [1]. While IMDG is known to induce a cell-mediated immune response, alum induces a humoral immune response. Bharat Biotech was the first company to use this kind of adjuvanation. It has been proven that Algel-IMDG-adjuvanted BBV152 is safe and immunogenic with the ability to induce Th1-biased T-cell responses. This would make COVAXIN® a potentially superior vaccine over other COVID-19 vaccines.

8.5 Beginning the Path to Progress: Preclinical Studies

Preclinical studies are essential in evaluating vaccine candidates' toxicity, immunogenicity and protective efficacy. Preclinical studies in mice, rats and rabbits were performed at RCC Laboratories, a subsidiary of Bharat Biotech in Hyderabad. The studies showed appropriate safety profiles and humoral and cell-mediated responses. Two protective efficacy studies were conducted in hamsters and non-human primates at the National Institute of Virology (NIV), Pune. The virus was rapidly cleared from the lower and upper respiratory tract in both studies, with no lung damage observed. The data from these studies indicate a strong immune response and protection against the virus, paving the way for human clinical trials. These data were published by the highly reputed Cell Press [2].

8.6 Human Clinical Trials at the Forefront

8.6.1 Phase 1

In July 2020, Bharat Biotech received approval from the Drug Controller General of India (DCGI) to initiate Phase 1 clinical trials. In a double-blinded, multicentre, randomised, controlled Phase 1 trial between 13 July and 20 July 2020, 827 healthy participants aged 18–55 were screened, and 375 were enrolled. Three hundred participants were randomly assigned to three vaccine groups (3 μg with Algel-IMDG, 6 μg with Algel-IMDG or 6 μg with Algel), and 75 participants were randomly assigned to the control group (Algel only). Two intramuscular doses of vaccine/control were administered on day 0 and day 14. These volunteers braved all the risks and participated in the clinical trials through the lockdown [2].

BBV152 led to tolerable safety outcomes and enhanced immune responses. These early clinical trials focused on safety and the vaccine's ability to trigger an immune response, and both Algel-IMDG formulations were selected for Phase 2 immunogenicity trials [3].

8.6.2 Phase 2

Between 5 and 12 September 2020, 380 participants were enrolled and randomly assigned to the aforementioned treatment groups. In the Phase 2 trial, BBV152 showed better reactogenicity and safety outcomes, and enhanced humoral and cell-mediated responses were observed in the 6 μg group compared to 3 μg. The 6 μg with Algel-IMDG formulation was selected for the Phase 3 trials [1].

8.6.3 Phase 3

Encouraged by the promising results from the early trials, COVAXIN® advanced to Phase 3 clinical trials, which began on 16 November 2020. Jointly initiated by BBIL and ICMR, this phase involved 25,800 participants across 25 sites and was conducted under strict regulatory oversight (Fig. 8.1). This data was instrumental in determining the vaccine's effectiveness against COVID-19.

By evaluating confirmed cases, COVAXIN® demonstrated 77.8% vaccine efficacy against symptomatic COVID-19 and 63.6% against asymptomatic COVID-19 [2].

COVAXIN® became the first COVID-19 vaccine to be tested and approved in adolescents and children in India. For Bharat Biotech, this level of safety and efficacy was paramount.

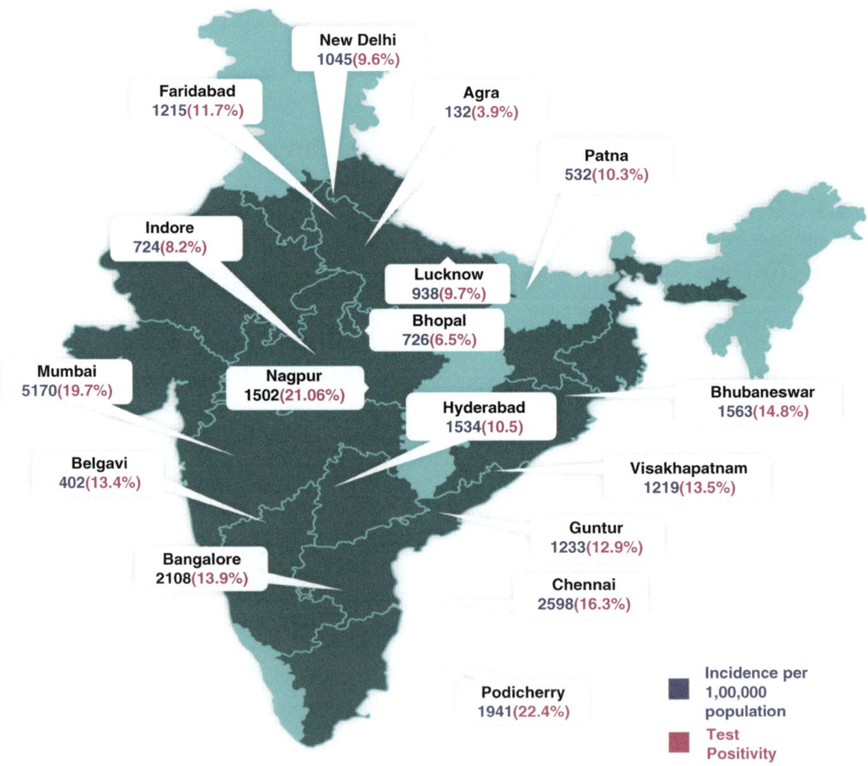

Fig. 8.1 Phase 3 efficacy trials were conducted in 25 sites on 25,800 subjects

8.7 Emergency Use Authorisation

In tandem with its scientific development, COVAXIN® underwent a rigorous regulatory process. The DCGI granted COVAXIN® Emergency Use Authorisation (EUA) in January 2021. This authorisation marked a turning point in the battle against COVID-19 in India, paving the way for the mass vaccination campaign. COVAXIN® also gained approvals in several countries and became integral to the global vaccine arsenal.

8.8 COVAXIN® Meets International Guidelines

All the studies reiterate global standards for the vaccine. The vaccine proved to be both safe and effective, meeting all the data expectations (2 months after Phase 3). Additionally, it neutralised most COVID variants, was consistently manufactured and followed up with a robust post-marketing survey, and trials were planned for immunocompromised and pregnant individuals. Its adherence to these guidelines has helped WHO and other international organisations to have faith in the quality of the vaccine.

8.9 Adaptations and Variants

The emergence of new variants of the SARS-CoV-2 virus added another layer of complexity to the COVID-19 pandemic. However, Bharat Biotech and India's COVAXIN® adapted to the evolving landscape. A study conducted by ICMR demonstrated COVAXIN®'s potent neutralisation and effectual antibody responses against the variants of concern. These included the strains Alpha (B.1.5.7), Kappa (B.1.617.1), Gamma (P2-B.1.1.28), Beta, and Delta (B.1.351 and B.1.617.2, respectively) [4]. The long-term protection through memory T- and B-cell responses was also studied. COVAXIN® induced robust T-cell responses that persisted for 6 months, even after antibody decline. Over the course of 12 months, T-cell responses in vaccinated individuals were carefully monitored, independent of whether they received a third dose. These T-cell responses played a pivotal role in generating a B-cell memory response upon re-exposure to the antigen, conferring robust and enduring long-term immunity to its recipients.

COVAXIN® served a multifaceted role extending beyond its capacity to prevent severe illness and hospitalisations. Its development illustrated India's steadfast commitment to combating COVID-19 through innovative and indigenous solutions. The effective public–private partnership of Bharat Biotech with the ICMR proved to be a winning recipe. COVAXIN® opened to the worldwide immunisation effort after receiving regulatory clearances and emergency use authorisations. The hope it generated in the gloom of the pandemic increased the belief of common people in the capabilities of its scientific community significantly.

Bharat Biotech believed that the pandemic has been global, and so should the vaccine. As the 'Pharmacy capital' of the world, India clung to 'Made in India, Made for the World' and pursued the efforts of manufacturing COVID-19 vaccines in numerous countries. Amidst the daunting challenge of delivering safe and economically viable vaccines worldwide, India shined as a paragon of excellence, steadfastly living up to its esteemed reputation.

8.10 Expanding Manufacturing Capacities

To meet the supply requirements for the Indian population and to support the Government of India's (GOI) Immunisation Program, additional facilities at Malur in Karnataka and Ankaleshwar in Gujarat were also activated to augment capacities. Whole-virion-inactivated vaccines require handling live viruses and require multiple levels of containment and purification. Such high purification levels are highly susceptible to losses, but the result was a purified and safe vaccine. Bharat Biotech agreed to transfer COVAXIN® technology to four external companies, namely Indian Immunologicals, BIBCOL, Haffkine Institute, and Hester Life Sciences. Ultimately, Bharat Biotech supplied an abundance of ~350 million vaccines.

8.11 Turning Challenges into Triumphs

The journey of COVAXIN® was not without its share of challenges. Questions were raised about the lack of publicly available data during the early stages. Bharat addressed these concerns through transparency, data sharing and commitment to rigorous scientific processes. The challenges ultimately served as catalysts for increased scrutiny in the global vaccine development process.

8.12 Equalising Health, a Dose at a Time

Bharat Biotech's mission was to make the vaccine access equitable, especially for low- and middle-income countries. This goal represents a significant step towards addressing global health disparities and promoting well-being. It not only supports economic growth and helps prevent the spread of deadly diseases but also aligns humanitarian values with the human spirit. Bharat has succeeded by collaborating with WHO, GAVI, UNICEF, and other international organisations.

8.13 A Better Route: Intranasal Vaccines

The end of the pandemic did not depend on making a vaccine alone. It depended on creating the most efficient and productive vaccine. After the success of COVAXIN®, Bharat Biotech International Ltd. (BBIL) launched the world's first intranasal vaccine, iNCOVACC® (BBV154), which can be administered as a two-dose regimen for primary vaccination and as a heterologous booster.

Formulations that involve entry into the body that the virus takes—the nose and mouth, or mucosal routes—offer degrees of protection their injectable counterparts may never achieve. There has also been a marked decrease in the capacity for intramuscular vaccinations and booster shots to block infections and transmissions. This has spotlighted the allure of nasal vaccines. They have evident ease in administration, are non-invasive and illicit a broad immune response. With the ability to achieve mucosal immunity, a nasal vaccine arrests and neutralises the virus before it can drop down to the lungs and establish an infection. The most significant advantage is the way it resembles a natural infection.

Higher antibody levels make the average person less likely to get sick from the virus. Research suggests that antibodies from vaccines decline over time. Booster shots are advantageous in that they can increase antibody levels up to 10 times. These are especially important in people with weak immune systems—including those who have cancer, HIV, co-morbidities, are taking medications or have had organ transplants.

After studying the advantages of intranasal over intramuscular vaccines, scientists from the University of Alabama have established that the ideal vaccination strategy may use an intramuscular vaccine, followed by an intranasal booster to elicit a long-lived immune response. This includes systemic IgG (antibodies found in the blood) and mucosal IgA (antibodies found in the mucous membranes like the nose and mouth). The systemic IgG and a broad repertoire of memory B- and T-cells send memory B- and T-cells to the nasal passages and guide their differentiation (Fig. 8.2). This triggers mucosal protection, including IgA secretion and tissue-resident memory cells in the respiratory tract [5].

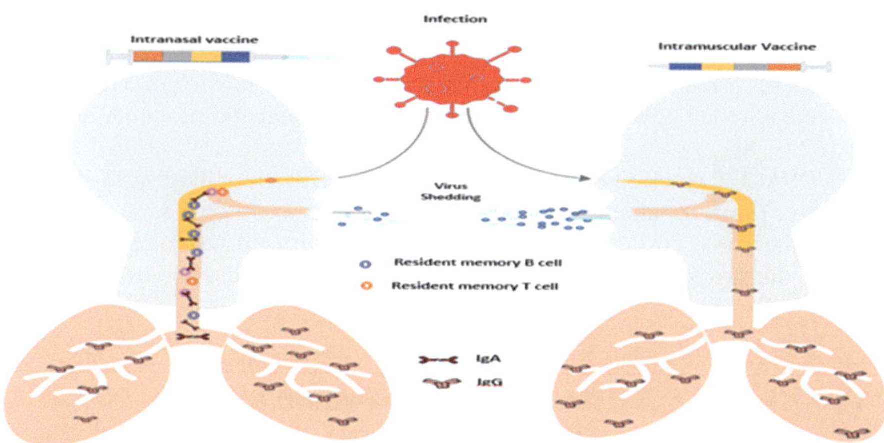

Fig. 8.2 Immune response elicited via an intranasal vaccine compared to an intramuscular vaccine

8.14 The Birth of iNCOVACC®

iNCOVACC®, also known as **BBV154,** was developed in partnership with Washington University, St. Louis, which had designed and developed the recombinant adenoviral vector construct and evaluated it in preclinical studies for toxicity and efficacy.

Bharat Biotech conducted product development related to preclinical safety evaluation, large-scale manufacturing scale-up, formulation and delivery device development and human clinical trials. When used as a heterologous booster following two doses of intramuscular shots of either COVAXIN® or COVISHIELD™, iNCOVACC® intranasal vaccine elicited a good protective immune response across SARS-CoV-2 variants, including Omicron.

In preclinical studies conducted in mice, rats, hamsters and rabbits, BBV154 elicited robust mucosal and systemic humoral and cell-mediated responses. In a SARS-CoV-2 challenge model using the highly susceptible K18-hACE2 transgenic mouse, one intranasal dose of BBV154 conferred superior immunity compared with one or two intramuscular immunisations with the same dose of the same vaccine. Furthermore, in studies conducted in K18-hACE2 transgenic mice, Syrian golden hamsters and rhesus macaques, an intranasal dose of BBV154 prevented upper and lower respiratory tract infections and inflammation due to SARS-CoV-2 [6].

8.14.1 Clinical Trials [1, 2 and 3]

Phase 1 and Phase 2 data of the clinical trials support the nasal vaccine's safety and durable immunogenic responses. After 6–12 months of the second phase, it met all the reactogenicity expectations. Following the demonstration of the safety and immunogenicity of BBV154 in humans, the interim immunogenicity and safety findings from a Phase 3, controlled, randomised, open-label trial of BBV154 were reported when administered as a homologous primary vaccination series comprising two doses given 4 weeks apart. Between 16 April and 4 June 2022, 3160 volunteers were enrolled, 2998 of whom received BBV154 and 165 of whom received BBV152 [7, 8].

iNCOVACC® received Emergency Use Approval (EUA) for those aged 18 years and above. It can be administered through the nose as a two-dose series, 4 weeks apart as the prime dose, and a booster can be administered heterologously. This makes it the first intranasal COVID-19 vaccine in the world to be used in these two ways.

8.15 Global Impact: Shaping a Better World Together

As we continue to navigate the challenges of the pandemic, the COVID-19 vaccines by Bharat Biotech stand as a symbol of scientific achievement, dedication and resilience. It serves as a reminder of the importance of collaboration, innovation and the

Fig. 8.3 QR code with the publications of COVAXIN® (scan to access)

relentless pursuit in the world of healthcare. The journey of COVAXIN® and iNCOVACC® is not just a success story for Bharat Biotech or India but a ray of hope for the world. The vaccines have played a crucial role in the global fight against COVID-19 and remain an inspiration to scientists and healthcare professionals worldwide (publications in the QR code in Fig. 8.3).

The 'bench-to-bedside' journey of a wholly indigenous COVID-19 vaccine in less than 8 months showcases the immense strength of *Atmanirbhar Bharat* (self-reliant India) to fight the odds and stand tall in the community. The epic saga of COVAXIN® by Bharat Biotech is a testament to the power of science, innovation and human determination. As we reflect on this incredible journey, we are reminded of the resilience of the human spirit and the unwavering commitment of scientists and healthcare professionals to safeguard our future. The two unsung heroes of this story, COVAXIN® and iNCOVACC®, have made Bharat Biotech the only company in the world to have made two COVID-19 vaccines indigenously. The story of Bharat Biotech continues to evolve, and it remains a beacon of hope in these challenging times we may confront in the future.

Competing Interest The authors have no conflicts of interest to declare that are relevant to the content of this chapter.

References

1. Ella R, Reddy S, Jogdand H, Sarangi V, Ganneru B, Prasad S, Das D, Raju D, Praturi U, Sapkal G, Yadav P. Safety and immunogenicity of an inactivated SARS-COV-2 vaccine, BBV152: interim results from a double-blind, randomised, multicentre, phase 2 trial, and 3-month follow-up of a double-blind, randomised phase 1 trial. Lancet Infect Dis. 2021;21:950–61.
2. Ella R, Reddy S, Blackwelder W, Potdar V, Yadav P, Sarangi V, Aileni VK, Kanungo S, Rai S, Reddy P, Verma S. Efficacy, safety, and lot-to-lot immunogenicity of an inactivated SARS COV-2 vaccine (BBV152): interim results of a randomised, double-blind, controlled, phase 3 trial. Lancet. 2021;398:2173–84.
3. Ganneru B, Jogdand H, Daram VK, Das D, Molugu NR, Prasad SD, Kannappa SV, Ella KM, Ravikrishnan R, Awasthi A, Jose J. Th1 skewed immune response of whole virion inactivated SARS COV 2 vaccine and its safety evaluation. iScience. 2021;23(24):102298.

4. Yadav PD, Sahay RR, Sapkal G, Nyayanit D, Shete AM, Deshpande G, Patil DY, Gupta N, Kumar S, Abraham P, Panda S. Comparable neutralization of SARS-COV-2 Delta AY.1 and Delta. J Travel Med. 2021:28–9.
5. Lund FE, Randall TD. Scent of a vaccine. Science. 2021;373:397–9.
6. Hassan AO, Shrihari S, Gorman MJ, Ying B, Yuan D, Raju S, Chen RE, Dmitriev IP, Kashentseva E, Adams LJ, Mann C. An intranasal vaccine durably protects against SARS-CoV-2 variants in mice. Cell Rep. 2021;36:109452.
7. Singh C, Verma S, Reddy P, Diamond MS, Curiel DT, Patel C, Jain MK, Redkar SV, Bhate AS, Gundappa V, Konatham R. Phase III Pivotal comparative clinical trial of intranasal (iNCOVACC) and intramuscular COVID-19 vaccine (Covaxin). npj Vaccines. 2023;8:125.
8. Ella R, Vadrevu KM, Jogdand H, Prasad S, Reddy S, Sarangi V, Ganneru B, Sapkal G, Yadav P, Abraham P, Panda S. Safety and immunogenicity of an inactivated SARS-CoV-2 vaccine, BBV152: a double-blind, randomised, phase 1 trial. Lancet Infect Dis. 2021;21:1473–3099.

9 Make in India Vaccines: A True Story of Atmanirbhar Bharat—Journey of COVID-19 Vaccines at Serum Institute of India

Adar C. Poonawalla, Umesh Shaligram, Sunil Gairola, Prasad S. Kulkarni, Bhagwat Gunale, Dhananjay Kapse, Chetanraj Bhamare, Umesh Shirsavkar, Amar Shrivastava, Manish Gupta, Manish Gautam, and Samir Naik

9.1 About SIIPL

Serum Institute of India Pvt. Ltd. (SIIPL) is the world's largest vaccine manufacturer by the number of doses distributed globally (more than 1.5 billion doses every year), which includes vaccines against Polio, Diphtheria, Tetanus, Pertussis, HIB, BCG, r-Hepatitis B, Measles, Mumps, Rubella, Pneumococcus, and COVID-19. It is estimated that about 65% of the children in the world receive at least one vaccine manufactured by SIIPL. Most of the SIIPL vaccines are prequalified by the World Health Organization (WHO) and are being used in around 170 countries across the globe, thus saving millions of lives.

SIIPL was founded in 1966 by Dr. Cyrus Poonawalla with the aim of manufacturing life-saving biologicals at affordable prices and in large quantities.

9.2 Introduction

The COVID-19 outbreak began in late 2019, in Wuhan, China. Subsequently, on March 11, 2020, the World Health Organization (WHO) declared the novel coronavirus (COVID-19) outbreak as a global pandemic [1]. Governments around the world began implementing various measures to control the spread of the virus including lockdowns, travel restrictions, and social distancing. In order to start vaccine development, the first need was to assess different technology platforms,

A. C. Poonawalla · U. Shaligram (✉) · S. Gairola · P. S. Kulkarni · B. Gunale · D. Kapse
C. Bhamare · U. Shirsavkar · A. Shrivastava · M. Gupta · M. Gautam · S. Naik
Serum Institute of India Pvt. Ltd., Pune, India
e-mail: umesh.shaligram@seruminstitute.com

© The Editor(s) (if applicable) and The Author(s), under exclusive license to Springer Nature Singapore Pte Ltd. 2025
R. Swarup et al. (eds.), *At War with the Single Strand*,
https://doi.org/10.1007/978-981-96-8122-8_9

considering the requirement for a large-scale manufacturing and vaccine delivery within a short turnaround time.

SIIPL started working on six different technologies for COVID-19 vaccine development in 2020. In a record time, we were able to deliver two commercial vaccines to the world. Here, we give the story behind SIIPL's development of COVID-19 vaccines (and therapeutics) during the pandemic. Due to these efforts, we were able to deliver more than 2.2 billion doses of COVID-19 vaccines, and out of this, more than 1.749 billion doses of Covishield were distributed in India alone.

9.3 Candidate Vaccines

We approached our existing collaborators across the globe as it was easily possible to get information regarding their product development for SARS-Cov-2. Legally as well, it was simpler to amend the ongoing agreements with the existing partners with the addition of new product development.

The first such partner was Oxford University with whom we were already working on the R21 malaria vaccine. They had made and clinically tested Ebola, chikungunya, malaria vaccines using a nonreplicating chimpanzee adenovirus-based platform (ChAdOx1). Thus, a reliable platform was already ready with them. Oxford had already initiated the development of COVID-19 vaccine by using this platform. After the release of the genetic sequence of COVID-19 virus by China on January 12, 2020, the Oxford team started designing ChAdOx1-based vaccine containing the genetic sequence of the spike protein of SARS-CoV-2. The vaccine trains the immune system to recognize and fight the virus if it is encountered in the future. The vaccine, ChAdOx1 nCoV-19, was called as AZD122 by AstraZeneca with the brand name Vaxzevria.

The second partner was Novavax due to our existing relations for Matrix-M which is used in R21 Malaria vaccine. Novavax used a classical spike protein-based technology along with the Matrix-M adjuvant. This vaccine, NVX-CoV2373, was made by Novavax at multiple places under the brand name of Nuvaxovid. The same vaccine was eventually made by SII in Pune with the brand name Covovax.

Novavax had earlier made RSV and influenza vaccines using this technology, and both were found safe and immunogenic in clinical studies.

Apart from this, there were five other approaches that were available to us like receptor binding domain (RBD) vaccine, measles virus vector-based vaccine (Themis), human adenovirus vector-based vaccine (Gamaleya), live attenuated vaccine (Codagenix), and mRNA vaccine.

There were lot of internal discussions at SIIPL about the best technology for making high volumes of vaccine in a short time. Based on our scientific analysis, we realized that most advanced products would likely be the vaccines being developed by Oxford and Novavax.

9.3.1 Covishield/Vaxzevria

SII had many discussions with Oxford about their technology and were ready for a technology transfer and eventual manufacturing. However, in April 2020, the University of Oxford entered into a partnership with AstraZeneca (AZ) that gave AZ global rights of the vaccine. Certainly, this was a shock to SII, though Oxford colleagues were helpful. As per their advice, SII approached AZ and eventually got into an agreement only for low- and middle-income countries (LMICs). While AZ made Vaxzevria in their UK plant, SII made the same vaccine in the Pune plant (later named as Covishield).

One critical aspect was manufacturing large quantities of the vaccine in the shortest possible time. In India, one cannot manufacture commercial batches of a new vaccine till a manufacturing and marketing license is given by the Indian regulator—the Drugs Controller General of India (DCGI) based on the Phase 3 clinical trial results. In a nutshell, after you receive the license, you start manufacturing the bulk which is eventually filled and finished and sent to the national control laboratory. After their release, the batch is brought into the market. As one can imagine, this process takes at least 4 to 5 months.

If India was to fight COVID-19 with the huge population, we cannot wait for manufacturing to start till the manufacturing license is received, and we needed to allow manufacturing when the product is in Phase 3 trials (even before we knew whether it is working or not). This involved a huge commercial risk for SII. If the product had failed in the Phase 3 trial, all the investment would have been lost. Thankfully, the Government of India agreed and issued an interim license which helped SII manufacture at risk and stock-pile the product even before the Phase 2/3 trial completion.

9.4 Clinical Trials

The clinical trials were a critical aspect in the licensure process. Vaxzevria was to undergo large Phase 3 efficacy trials in UK and other countries. We felt that another efficacy trial in India was not necessary for Covishield as both the vaccines were essentially identical except for the manufacturing location. Therefore, for Indian licensure, we proposed an immunological bridging trial of Covishield against Vaxzevria while comparing its safety with a placebo which would save enormous amount of time, money, and resources. The Indian regulators (DCGI) agreed with our proposal.

The clinical trials of Vaxzevria began in April 2020, with phase 1 and 2 trials involving around 1000 participants in the UK. The trials showed that the vaccine was safe and well-tolerated and that it produced a strong immune response in the participants [2]. Phase 2/3 trials of the vaccine began in the UK, Brazil, and South Africa in August 2020 and involved more than 20,000 participants. The trials showed that the vaccine was highly effective in preventing COVID-19, with an efficacy rate of around 64% after a single dose and up to 90% after two doses [3].

SII began conducting Phase 2 /3 immunobridging study of Covishield in August 2020. The Phase 2/3 clinical trials involved 1600 adults in India and were designed to evaluate the safety and immunogenicity of the vaccine. An independent data safety monitoring board (DSMB) comprising physicians and a biostatistician was constituted to keep an oversight on the safety of the participants. The trial met noninferiority to Vaxzevria in terms of immune response and was safe. Further, there were no severe cases of COVID-19 in the vaccine group throughout the study [4].

We faced many challenges during the trial. First, there was an overwhelming response to take part in the trial. However, some participants posted their pictures in social media which affected their confidentiality. We discovered the same situation in other countries as well. Though people were willing to take part, the commute of participants to the study sites was difficult. Also because of strict inclusion/exclusion criteria, the recruitment was a bit slow.

Every development in the trial was being reported in the media. The recruitment status was daily discussed in the public domain. Regulatory permissions were widely publicized. It was clear that the whole trial was under a big media glare.

In September 2020, the UK trial of Vaxzevria was halted because of a case of transverse myelitis in a participant. This became a big news in the Indian media, and as a result, we were asked to halt our trial as well. In a few days, the event was found to have no connection with the vaccine, and our DSMB, followed by DCGI gave us a go ahead to resume the trial.

In October 2020, a participant in Chennai reported a neurological event which resolved within 3 weeks and was not causally related to the study products. However, he went to the press, and the event got an adverse media publicity. He also eventually sued SII for compensation. However, the case did not stand in the court.

While the study follow-up was going on, Covishield received EUA, and the immunization programme started on January 16, 2021 (Fig. 9.1). As a result, many volunteers started demanding unblinding so that if they had received a placebo, they could opt for the vaccination in government program. We took special permission from the regulators to allow the same.

After the program started, the vaccinees started getting vaccination certificates from the government. These certificates became mandatory for travel and visits to many places like malls and offices. The trial participants could not get those certificates because they had received vaccines in the trial. This created a major inconvenience to the participants. We approached the government with this issue, and they issued vaccination certificates to all trial participants as per their request.

Fig. 9.1 At the time of the first shipment of Covishield vaccine on January 12, 2021

9.5 Technology Transfer

In a typical technology transfer, there are physical meetings between the two parties where they exchange the knowledge of process/analysis and establish the proof of concept (POC). But in this case, entire technology transfer was done through online meetings. Without hands-on training, SIIPL initiated manufacturing campaign based on the information from technology transfer documents, online meetings, and discussions. During development, there were frequent changes in the process. Finally, SII managed to develop and optimize the complete process with limited resources. Developmental gaps were bridged seamlessly by accessing internal expertise to overcome limited technology knowledge.

9.6 Challenges in Manufacturing

There were serious challenges in the early days of manufacturing. The first and biggest challenge was working during the strict lock-down. Access to people and resources was extremely limited, and there was a fear factor among people that affected their willingness to join for work. Hiring the new skilled staff was difficult, as most people did not want to change jobs during the pandemic. Yet, SII recruited around 2000 employees during this period, of which around 90% were fresh graduates. The existing employees had to train the new recruits along with managing the ongoing developmental and manufacturing activities.

There were travel restrictions, and commercial flights were not operating. This ruled out any domestic or international travel.

Raw material supply had severe constraints from western countries' manufacturers. The PMO helped us to get clearances so that foreigners could travel to India and install large disposable reactors and filling lines. The SII project team ensured that within a few months, facilities were ready and equipment were installed. Some of the foreigners became COVID-positive, but our Medical Department provided them necessary medical care.

Initially, we had ordered the raw material considering a certain quantity of doses to be made; however, as the pandemic raged, there was a sudden high demand of the vaccine. Moreover, the demand was unpredictable, and the exact quantity was still not known. This put us in a difficult position about ordering the raw materials.

In the initial days, we were also struggling to standardize the method of manufacturing. We had no experience of working on the Ad-vector platform, but with sheer determination and hardwork, we overcame this issue and with round the clock work, we could stock-pile 90 million doses. We, for the first time, used the technology of a single-use manufacturing platform and that significantly increased the efficiency and flexibility of manufacturing, enabling rapid production and distribution of vaccines during the pandemic.

9.7 Traditional vs. Single-Use Vaccine Manufacturing

In traditional vaccine manufacturing, production facilities use large stainless-steel bioreactors to grow cells and produce the vaccine. The bioreactors must be cleaned and sterilized after every batch of vaccine is produced, which can be time-consuming and labor-intensive. Additionally, the use of stainless-steel bioreactors requires a significant amount of space, which can be a constraint for some production facilities.

A single-use manufacturing platform involves the use of disposable plastic bags and tubings instead of stainless-steel bioreactors. This allows for faster and more flexible production, as the disposable components can be quickly assembled and disassembled without the need for cleaning and sterilization (Fig. 9.2).

Single-use components are also easier to transport and store, as they are lighter and require less space. The development of a single-use manufacturing platform has enabled faster, more flexible, and more efficient manufacturing of vaccines such as Covishield. The rapid deployment of millions of doses of Covishield across the world is a testament to the impact that new technologies can have on global health.

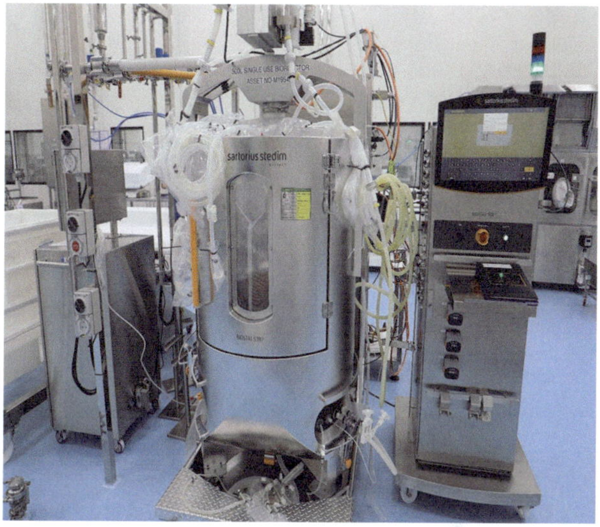

Fig. 9.2 HEK 293 cell seed manufacturing for Covishield at Manjari facility

9.8 Raw Material and Consumables

Most of the consumables were single-use materials and were available from a single vendor which limited the quantity. New multiple vendors were developed, established, and qualified for the process. The BOM was modified multiple times like chromatography membrane, TFF cassettes, mixing and holding bags, DS storage bags, connectors, etc. Huge shortage of critical raw materials like media, feed, and endonucleases was mitigated through continuous and extensive follow-up with vendors and buyers. Common raw materials like Tris, sucrose, and other excipients were in short supply globally; therefore, multiple alternate sources were introduced in the process, and multiple times suitability was established as required. Continuous supply chain management hurdles were faced but eventually overcome.

9.9 Process Development

COVID 19 cases among employees resulted in the shortage of trained manpower. New facilities were set up and close-loop operations were introduced to minimize the cross-contamination. To meet the high dose requirement, scale-up of 2000 L, 4000 L, and 6000 L facilities was established within a short time (Fig. 9.3).

Fig. 9.3 Covishield vaccine manufacturing facility using six single-use 4KL bioreactors at Manjari facility

9.10 Commercial Studies

Services with respect to qualifications, maintenance, repair, etc. from vendors were delayed due to Covid-19 protocols and travel restrictions. Overall shipments and outsourced sample analysis studies got hampered due to travel restrictions. Most of the critical equipment like bioreactors, chromatography systems, and TFF systems were procured without the Factory Acceptance Test; as a result, difficulties were faced during operations.

9.11 Manufacturing Challenges and Vaccine Quality

Product quality is the key for any vaccine for its safety and efficacy. We implemented complete close-loop operations by using sterile/Gamma-irradiated process assemblies to avoid any open operations and cross-contaminations. This was essential in order to reduce potential batch failures and increase the production capacity to its maximum. Additionally, to monitor the integrity of virus, we implemented NGS-based whole-genome sequencing analysis as a part of additional characterization. First time, we developed and implemented an NGS-based adventitious agent analysis for our cell banks, and virus stock and regulators accepted this high-throughput data against traditional analysis. This resulted in a significant reduction of time for batch release. Now, regulatory bodies across the globe have started accepting NGS data and are in the process of drafting new guidelines. We also monitored batches for TEM analysis and Cryo-EM analysis to visualize the vaccine quality and the structure of virus.

Considering the demand, we converted a lot of our existing GMP areas for COVID-19 vaccine manufacturing, while we scaled up DS manufacturing at 4000 L scale with a unique fermenter (Fig. 9.3). This was not being used by anyone for COVID-19 vaccine manufacturing, and as a result, bioreactor bags and others things were easily available from Ireland. Sometimes, private jets were sent to get the bags. Because of all these steps, we could make 300 million doses per month.

9.12 Fill Finish Capabilities

The SEZ-5 GF fill finish line was qualified for DTP group of vaccines and was diverted for COVID-19 production. This was an isolator line which can run on campaign basis to have a consistent and high output. Primary Packaging Materials (PPMs) such as vials and stoppers meant for other vaccines were diverted to COVID-19 vaccine manufacturing, and additional orders were placed to the suppliers. To install filling lines, project and engineering teams worked together with the operational team to complete the qualification and made the filling line operational. Three additional machines were put in operation which successfully filled more than two billion doses.

9.13 Process Scale-Up

Initially, we manufactured Covishield vaccine at a 2000 L scale; later, we set up a new facility with a capacity of 4000 L bioreactors (Fig. 9.3). Manufacturing in this facility was started in April 2021, and by November 2021, it delivered more than billion doses.

9.14 Regulatory Approvals

Since both Oxford and Novavax were recombinant products, we required the Review Committee on Genetic Manipulation (RCGM) and the Department of Biotechnology (DBT) permissions before the development. On March 19, 2020, SII team met in Delhi with the then DBT Secretary, Renu Swarup, and RCGM team members. RCGM gave an approval for BSL-1 classification of Ad-vector.

The ChAdOx1 nCoV19 vaccine (AZD1222, AstraZeneca) received emergency use approval from the UK Medicines and Healthcare Products Regulatory Agency (MHRA) in December 2020, and Covishield (ChAdOx1 nCoV19) was granted emergency use authorization (EUA) from the DCGI on January 3, 2021. It was also approved for emergency use in several other countries, including Mexico, Brazil, and Argentina. The approval of Covishield for emergency use was a major milestone in the fight against COVID-19 (Fig. 9.4). On January 16, 2021, the Government of India launched the world's largest Covid-19 immunization program, initially targeting high-risk groups, such as healthcare workers and elderly people (Figs. 9.1 and 9.5).

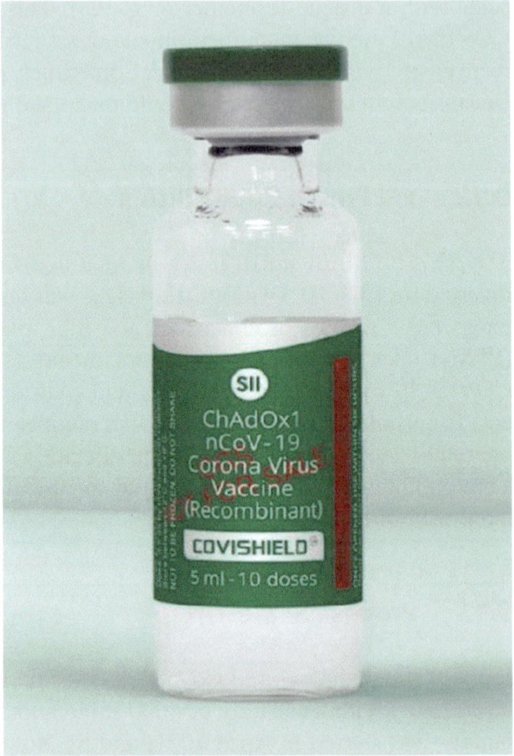

Fig. 9.4 Covishield 10 dose vial

Fig. 9.5 First shipment of Covishield vaccine to UK government. This was the first vaccine supplied to UK from India on March 4, 2021

Subsequently, Covishield received emergency use listing/EUA from several international authorities including ANVISA (January 20, 2021), WHO (February 15, 2021), MHRA (February 23, 2021), Health Canada (February 26, 2021), etc. Covishield (ChAdOx1 nCoV19) is the first vaccine made in India (and in any developing country) that was supplied to developed countries (Figs. 9.6 and 9.7). Finally, full marketing approval was granted in India on January 27, 2022.

Fig. 9.6 Covishield Serum Institute of India manufactured vaccine in India and distributed to 98 countries across the globe

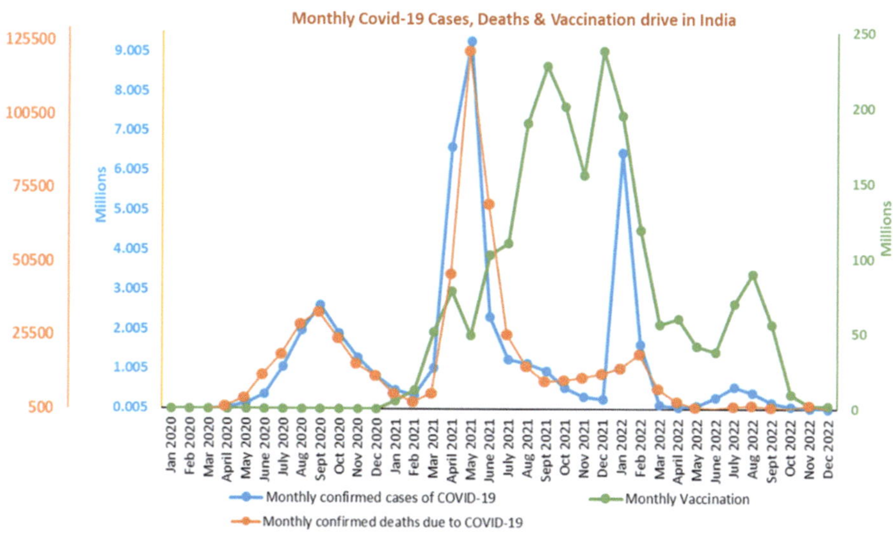

Fig. 9.7 Graphical representation of monthly Covid-19 cases from January 2020 to December 2022 and the vaccination drive in India from January 2021 to December 2022. As the vaccination drive accelerated, the Covid-19 cases started declining. It also shows the monthly confirmed death trend for the same duration

9.15 Government Support

Throughout this journey, the government of India and their offices were highly supportive. The DCGI gave timely approvals without compromising on the quality of the product. The Indian Council of Medical Research (ICMR) supported clinical trials. The DBT gave permissions on a fast-track basis. The foreign ministry gave emergency visas to foreign collaborators and workers that helped in installing the new machines.

9.16 Vaccine Supply and Impact

Initially, there was a lukewarm response to vaccination, even by the high-risk populations. Many people thought that the pandemic was over as the case numbers were subdued and there were few hospitalizations. The fear was gone. In fact, during the initial round, it was difficult to get 10 people for using a 10-dose vial (Fig. 9.4). As a result, many doses were wasted.

In March 2021, the cases started rising. A new variant, Delta started creating havoc in India (and in many other countries). Within few days, the hospitals were full, oxygen was in short supply, ICU beds were not available, and Remdisivir was difficult to get. The death toll started rising. There was panic all over the country. As a result, the demand for the vaccine went through the roof. There was a huge pressure for the vaccine doses from all over the world.

SII used their new facilities to the fullest extent with people working round the clock. Staff worked with a passion. Many of them had lost their near and dear ones. All that was paid off, and SII started supplying huge quantities to the government who ran a very efficient program across the country. As the vaccination increased, the virus started retreating, and the hospitalizations and deaths became rare. The lockdowns were lifted, the routine life started, and economy started coming back to normal.

SII supplied two billion doses to the Indian government, private hospitals in India, and to other countries. Covishield, along with Vaxzevria, became the most commonly used COVID-19 vaccine in the world (Fig. 9.7).

9.16.1 Pharmacovigilance

Traditional vaccine development takes around 7 to 10 years. The COVID-19 vaccines were developed, tested, underwent clinical trials, and got approvals in just around 9 months. That was a phenomenal speed. This was possible because of several factors: the vaccine platforms were already available, the regulatory permissions were fast, the trial recruitments were fast, for efficacy assessments, the COVID-19 cases were accrued quickly because of high attack rates, etc. However, this fast development also gave rise to apprehensions that whether all safety parameters were really checked diligently during these vaccines' development. As a result, the vaccines received emergency use approvals with stringent marketing conditions. The regulators needed the review of safety data collected from ongoing trials as well as post-authorization use every 15 days in the form of safety summary reports.

The vaccine distribution and administration were controlled by the government. SII put in place a robust pharmacovigilance system to collect all the safety-related information and to detect any safety issues/signals. A separate fact sheet was prepared and was provided with all vaccine vials going to the field. This fact sheet addressed all the common safety concerns about the vaccine. The fact sheet also encouraged people to report adverse events to our 24/7 toll-free number, on our dedicated email address, or on the postal address. Our toll-free number, email ID, and postal address were given in the fact sheet as well as on our website. All adverse events published in newspapers/social media/online stories/medical and scientific journals were looked for. All the events reported from these different sources were logged in our system. Each event was evaluated, and an assessment report was prepared and submitted to the Pharmacovigilance Programme of India (PvPI) and to the DCGI.

We received thousands of emails, phone calls, letters, publications asking medical enquiries, and reporting adverse events. For no other vaccines, we have received so many reports. There was a lot of misinformation floating on the social media that fueled rumors about the safety/efficacy of the vaccine. We hired more people, and our vendors also hired more staff to meet the huge case load and to comply with the regulatory requirements.

All types of complaints, medical enquiries, and adverse events were received. Each adverse event was recorded, evaluated, replied, and reported. Most were either coincidental or anxiety reactions or transient vaccine-related reactions. A very rare adverse event of "thrombosis with thrombocytopenia syndrome (TTS)" was reported with the use of ChAdOx1 nCoV-19 vaccine from different countries. Since there was no information available in the initial period, these events were not treated appropriately, and as a result, few deaths occurred. After the event was identified and appropriate information for the diagnosis and management was disseminated, the deaths could be prevented with a correct treatment. The incidence of the event was less than 1 in 100,000 doses globally and 1 in 70 million doses in India.

9.17 Vaccine Impact

In the first wave in 2020, the most commonly affected groups were frontline workers like doctors, nurses, hospital staff, and police as well as senior citizens. There were a lot of hospitalizations and deaths in these groups.

In January 2021, Government of India started the vaccination drive of the above groups using Covishield vaccine. When the deadly Delta wave started in March 2021, most of these people were already vaccinated. As a result, there were fewer cases/hospitalizations/deaths in these people (Fig. 9.8). Moreover, since healthcare workers were already protected, hospitals could run smoothly, and as a result they were able to treat millions of patients. As vaccination was not yet started for the age group of 18–45 years, this population was most affected during the Delta wave.

Fig. 9.8 Covishield vaccine manufacturing facility using six single-use 6KL bioreactors at Manjari facility

As the vaccination progressed, the cases receded. Later, the Omicron wave affected US and European countries with vaccine breakthrough infections and mortality in unvaccinated population. In India, before the third wave of Omicron, around 1.75 billion doses were used, beginning from the inception of vaccination drive, and among these, ~89% doses were of Covishield. As a result, hospitalization and mortality were negligible during the Omicron wave (Fig. 9.8).

About 11.2 million doses of Covishield were also supplied to over 96 countries under the *Vaccine Maitri* initiative of the Government of India. AstraZeneca made contracts with more than 30+ contract manufacturing organizations (CMOs), including SII. Altogether, AZ made a total of 3.2 billion doses by early 2022, of which SII alone made 2 billion doses.

Covishield was also used as a homologous booster in people who had received primary immunization with Covishield. Around 940 million doses were used as a booster and precaution dose.

This was possible because of the huge commercial risks taken by Dr. Cyrus Poonawalla and Mr. Adar Poonawalla. In fact, at the later phase of the pandemic, when the vaccine uptake went downhill, SIIPL had to destroy 125 million doses worth USD 276 million.

9.17.1 Recombinant BCG Vaccine

When the pandemic struck, there were no covid vaccines available nor was there any clarity on when they would become available. As a result, it was decided to pursue a quick ready-to-go product. It is well known that BCG, that is commonly used to prevent tuberculosis, also activates innate immunity that helps in the reduction of other infections. This is something called as nonspecific effects of BCG vaccine.

SIIPL is developing VPM1002 (recombinant BCG) vaccine which is a modified version of the existing BCG vaccine. It was felt that recombinant BCG, which was in advanced stages of clinical development for the prevention of TB, could be repurposed for the prevention of severe respiratory disease, including those caused by COVID-19. Thus, the world's first vaccine to go into clinics for the prevention of severe COVID-19 was recombinant BCG vaccine.

We conducted two trials, one in Germany and one in India. The study in Germany showed that severe respiratory disease was prevented by the vaccine.

9.17.2 Covovax/ Nuvaxovid Vaccine

This vaccine was developed in collaboration with Novavax, USA. Novavax vaccine (***NVX***-CoV2373) is a protein-based vaccine manufactured using the baculovirus–insect cell line expression system. The baculovirus containing the recombinant antigen gene is used to infect cells (called Sf9 cells). The baculovirus multiplies (replicates) inside these cells. The spike protein expressed in insect cells is purified through multiple purification steps. The vaccine is adjuvanted with Matrix M (saponin-based adjuvant). The vaccine made by Novavax was branded as Nuvaxovid, while the one made by SII is called Covovax.

SIIPL executed a contract with Novavax to import the DS manufactured at their site, manufacture fill finish at SIIPL, and further supply them the drug product (DP). Eventually, SII also manufactured the DS. The Novavax DS and SII DP product were used in the Phase 2 trial, while Phase 3 trial was completed using DS and DP manufactured at SIIPL. The trial demonstrated comparable safety and met noninferiority in terms of the immune response of Covovax against Nuvaxovid. Novavax vaccine manufacturing was probably the toughest technology (based on baculovirus/insect cell line platform) to understand and further develop and scale up quickly.

Novavax had their own manufacturing unit in Praha (Czech Republic), and they also had contracts with more than eight CMOs, including SII to manufacture the vaccine. However, SIIPL was the only one that successfully developed and scaled up the manufacturing. For this, our ongoing recombinant BCG production was stopped, and its facility was converted into Covovax facility.

Initially the development and manufacturing happened at 200 L scale from which the product went to the Indian Phase 2/3 trial. Later, the scale-up was done at 2000 L scale. We also added up an additional 6000 L facility (Fig. 9), considering the demand of about five to six billion doses from COVAX and GAVI (agencies which were procuring the vaccine from the vaccine manufacturers and supplying to the countries where the vaccine was required, mainly the lower middle-income countries).

SIIPL was the first site from where NVX-CoV2373 vaccine was approved by various regulatory agencies. The first approval was given by Indonesia, and SIIPL supplied nine million doses of Covovax to Indonesia. Subsequently, we got approval from the agencies such as WHO, DCG(I), Food and Drug Administration (FDA), European Medicines Agency (EMA), and Therapeutic Goods Administration (TGA) to name a few.

NVX-CoV2373 was the first vaccine manufactured in any developing country that was supplied to US, Europe, Australia, etc. The FDA approval was obtained in July 2022. Recently, SIIPL has developed the XBB.1.15 vaccine and supplied the same to US.

9.17.3 Receptor-Binding Domain (RBD) Vaccine

This was another candidate vaccine that we worked on. We decorated the surface of Hepatitis-B VLP with the receptor-binding domain (RBD) of SARS CovV-2 by using the Spytag-Spycatcher technology. We standardized the method of manufacturing and completed a Phase 1/2 trial in Australia. This vaccine also demonstrated good results. Meanwhile, we got two vaccines (Covishield and Covovax) licensed, and therefore, we decided to discontinue its further development.

9.17.4 COVID-19 Antiserum

In parallel, we developed a therapeutic antiserum using horses by immunizing horses with the RBD vaccine to generate sera by the classical method of manufacturing antisera. We felt this was the only cost-effective therapeutic option for LMICs as it is difficult to manufacture monoclonals or cocktails of monoclonals cost-effectively. The antiserum contained Fab2 against the RBD of SARS COV2 virus-enveloped protein. This anti-RBD Fab2 executes the antiviral activity against SARS COV2 virus by neutralization.

Acute dose and repeat dose toxicity studies in small animals were completed with the antiserum that showed no toxicity. Neutralization studies were conducted at the West Virginia University, USA, which showed that the antibodies can effectively neutralize the different variants of concern (VOCs) [5]. Subsequently, it was tested in a Phase I clinical trial in 40 adult patients with moderate COVID-19 disease. The antiserum was found safe. Finally, we stopped the further development due to commercial reasons, but this technology can be easily made available for the next pandemic.

9.18 Pandemic Preparedness Facility

Going ahead, SIIPL is also constructing its own pandemic preparedness facility with an annual capacity of producing five to seven billion doses. It will help in catering a huge demand of vaccines in a very short time in case of any future pandemic.

This is an independent manufacturing building with four floors of 70,000 sq. feet each. The building would be equipped with its own infrastructure services to cater the manufacturing block. Two complete service floors are dedicated for four bulk manufacturing suits. These manufacturing suits will be capable of handling various technologies like mRNA, live attenuated viruses, inactivated viruses, subunit vaccine, and recombinant-based technologies using mammalian, bacterial, and yeast cell lines.

One dedicated manufacturing suit will be compliant to BSL-3 guidelines to handle live virus in case of pandemic. State-of-the-art fill finish facility would be a part of the building which can handle both liquid and freeze-dried products. There will be dedicated huge cold rooms for storing final products along with packing, and dispatch would be the additional feature of the building.

This facility has multiple bulk manufacturing facilities designed on both single-use and stainless-steel technology platforms that support the manufacturing of microbial and mammalian cell line products, both intracellular and extracellular types. The facilities have been designed for multiproduct manufacturing. The installed capacities are high that shall support not only regular vaccines manufacturing at SIIPL but also can be ramped up to manufacture other vaccines in case of a pandemic to meet high demands. This also includes a new facility designed for the manufacturing of vaccines on mRNA technology.

Fill finish capacity is also huge with three separate facilities with two liquid lines of 500 vials (2 mL)/min capacity and one lyophilization line of 500 vials/min capacity with two integrated lyophilizers of 40 sqm each. There is another facility with a pilot filling line for vials and pre-filled syringes (PFS) with a capacity of 100 vials and PFS/min. Apart from this, the facility has dedicated warehouse, packing, and dispatch areas to support the manufacturing and release of products.

9.19 Discussion

The COVID-19 pandemic was a global crisis that caused more than 771 million confirmed cases, including around 7 million deaths as of October 25, 2023. SIIPL worked on different approaches: general immunity boost with the recombinant BCG vaccine, therapeutic RBD antisera, and four different COVID-19 vaccine candidates. This multipronged approach was taken due to a lot of unknowns. This culminated into the successful story of supplying billions of doses of Covishield and Covovax across the globe which was possible because of the knowledge, skill set, and dedication of SII people.

SII made huge investments in building the facilities. We had to sacrifice the production of other products such as recombinant BCG vaccine, erythropoietin, human

Table 9.1 Country-wise vaccination data showing Indian vaccination drive is second in the world [7]

Country name	Vaccinated	Fully vaccinated	Total
World	**72%**	**67%**	**13,270,170,332**
Mainland China	94%	91%	34,491,077,000
India	75%	70%	2,206,348,628
United States	80%	68%	632,905,664
Brazil	90%	83%	484,995,271
Indonesia	75%	64%	444,303,130
Japan	83%	82%	381,675,627
Bangladesh	93%	84%	345,933,184
Pakistan	75%	64%	333,662,084
Vietnam	94%	89%	266,211,063
Mexico	78%	65%	225,063,079
Germany	78%	76%	192,102,093
Russia	61%	55%	185,672,750
Philippines	73%	68%	170,457,489
Iran	79%	71%	157,785,811

papilloma virus vaccine, acellular pertussis vaccine, and monoclonal antibodies. The net result was a supply of almost 2.2 billion doses across the world.

Our partner, AstraZeneca, worked with more than 30 CMOs and altogether supplied 3.2 billion doses, of which 2 billion doses were Covishield alone. Covishield is estimated to have saved around 4 million lives and prevented more than 32 million infections [6]. In Indian vaccine program, the SIIPL contribution was more than 80%.

Similarly, Novavax, another partner, was working on 10 plants including their own plant, but only the SIIPL plant was approved by US FDA, and we supplied 80 million doses. This is the first vaccine manufactured by any developing country manufacturer, which was supplied to USA, Europe, Australia, and New Zealand (Table 9.1).

Every possible solution was put into translation so that we could achieve a solution for this crisis. Since two products were successfully approved, we did not pursue the development of other approaches, but these platforms are now ready, and they can be useful for any future pandemics.

Competing Interest The authors have no conflicts of interest to declare that are relevant to the content of this chapter.

References

1. World Health Organisation (2020). WHO Director-General's Opening Remarks at the Media Briefing on COVID-19 – 11 March 2020. [online] World Health Organization. Available at: https://www.who.int/director-general/speeches/detail/who-director-general-s-opening-remarks-at-the-media-briefing-on-covid-19%2D%2D-11-march-2020.

2. Folegatti PM, Ewer KJ, Aley PK, Angus B, Becker S, Belij-Rammerstorfer S, Bellamy D, Bibi S, Bittaye M, Clutterbuck EA, Dold C. Safety and immunogenicity of the ChAdOx1 nCoV-19 vaccine against SARS-CoV-2: a preliminary report of a phase 1/2, single-blind, randomised controlled trial. Lancet. 2020;396(10249):467–78.
3. Ramasamy MN, Minassian AM, Ewer KJ, Flaxman AL, Folegatti PM, Owens DR, Voysey M, Aley PK, Angus B, Babbage G, Belij-Rammerstorfer S. Safety and immunogenicity of ChAdOx1 nCoV-19 vaccine administered in a prime-boost regimen in young and old adults (COV002): a single-blind, randomised, controlled, phase 2/3 trial. Lancet. 2020;396(10267):1979–93.
4. Kulkarni PS, Padmapriyadarsini C, Vekemans J, Bavdekar A, Gupta M, Kulkarni P, Garg BS, Gogtay NJ, Tambe M, Lalwani S, Singh K. A phase 2/3, participant-blind, observer-blind, randomised, controlled study to assess the safety and immunogenicity of SII-ChAdOx1 nCoV-19 (COVID-19 vaccine) in adults in India. EClinicalMedicine. 2021:42.
5. Barbier M, Lee KS, Vikharankar MS, Rajpathak SN, Kadam N, Wong TY, Russ BP, Cyphert HA, Miller OA, Rader NA, Cooper M. Passive immunization with equine RBD-specific Fab protects K18-hACE2-mice against Alpha or Beta variants of SARS-CoV-2. Front Immunol. 2022;13:948431.
6. Watson OJ, Barnsley G, Toor J, Hogan AB, Winskill P, Ghani AC. Global impact of the first year of COVID-19 vaccination: a mathematical modelling study. Lancet Infect Dis. 2022;22(9):1293–302.
7. Holder J. Tracking Coronavirus Vaccinations Around the World. The New York Times [online] 13 Mar. 2023. Available at: https://www.nytimes.com/interactive/2021/world/covid-vaccinations-tracker.html.

Make in India Vaccines: A True Story of Atmanirbhar Bharat—ZyCoV-D, the World's First Human DNA Vaccine for COVID-19

10

Kapil Maithal, Ravindra Mittal, Deven Parmar, Mukul Jain, and Pankaj R. Patel

10.1 Background

A new coronavirus, designated 2019-nCoV, was reported in Wuhan, Hubei Province of China, in late 2019/early January 2020 for the first time. Subsequently, several countries across Europe, North America, and Asia-Pacific started reporting cases of coronavirus infection (COVID-19). This led to WHO declaring COVID-19 as a global pandemic and the Central Government of India declaring the pandemic as a "notified disaster" under the Disaster Management Act, 2015. Many Indian manufacturers ventured into the development of vaccine(s) against COVID-19 either using in-house expertise or technology transfer from academic partners in India or abroad. On similar lines, Zydus Lifesciences ventured into the development of a COVID-19 vaccine using in-house capabilities on the highest priority for use by masses.

Considering the urgency, available literature, and infrastructure, a plasmid DNA vaccine targeting the spike protein of SARS-CoV-2 virus was considered to be a suitable option. It is noteworthy to mention that until that time, primarily DNA vaccines for other coronaviral diseases like SARS-CoV-1 and MERS-CoV had progressed successfully in clinical trials globally. Both MERS-CoV and SARS-CoV-1 DNA vaccines were found to be safe and well tolerated in humans. MERS-CoV DNA vaccine had a seroconversion rate of 94% in vaccinated volunteers, whereas the SARS-CoV-1 DNA vaccine induced antibody response in 80% subjects [6, 7].

K. Maithal · R. Mittal · D. Parmar · M. Jain · P. R. Patel (✉)
Vaccine Technology Center, Zydus Lifesciences Limited, Ahmedabad, India

Zydus Research Center, Clinical R & D, Zydus Lifesciences Limited, Ahmedabad, Gujarat, India

Zydus Lifesciences Limited, 'Zydus Corporate Park', Ahmedabad, India
e-mail: pankaj@zyduslife.com

© The Editor(s) (if applicable) and The Author(s), under exclusive license to Springer Nature Singapore Pte Ltd. 2025
R. Swarup et al. (eds.), *At War with the Single Strand*,
https://doi.org/10.1007/978-981-96-8122-8_10

The plasmid DNA vaccine platform had numerous potential advantages over traditional vaccines in terms of effectiveness, safety, and cost. Some of the major factors favoring this platform are as follows:

1. DNA vaccines ensure appropriate folding of the polypeptide and produce the antigen over long periods without the requirement of adjuvants. These host-synthesized antigens become the subject of immune surveillance in the context of both major histocompatibility complex class I (MHC I) and MHC II proteins of the vaccinated individual [5] and elicit both humoral and cellular arms of the adaptive immune response [2].
2. DNA vaccines also have excellent safety profile as they do not use microorganism(s), and therefore, there is no risk of reversion.
3. The platform is easily adaptable, wherein the antigen can be modified rapidly based on the new emerging variants. This will provide flexibility in programmatic implementation to switch over to newer vaccine candidates based on the emerging new variants.
4. DNA vaccines are usually quite stable and may be stored at ambient temperature for weeks/months besides being able to withstand multiple freeze thaws, a major challenge in mass vaccination associated with temperature excursions.
5. DNA vaccines do not have the problems often associated with vector-based immunity, such as poor immune response to target antigens following vaccination due to the pre-existing antibody titers to vectors resulting from natural infections from vectors such as adenoviruses, measles, and influenza viruses.

In spite of these advantages, until that time, only few veterinary DNA vaccines/therapeutics were approved globally and none for human use. These candidates were as follows:

1. West Nile DNA vaccine for Equines (Fort Dodge Animal Health, commercialized by Pfizer)
2. Infectious Hematopoietic Necrosis Virus vaccine for Salmon (Canadian Food Inspection Agency, Apex-IHN® produced by Novartis)
3. Therapeutic vaccine for cancer treatment in canine (USDA, ONCEPT™ produced by Merial)
4. Gene therapy for pigs (LifeTide® SW 5, produced by VGX Animal Health)

The major obstacle in the success of human DNA vaccine until that time was lack of good immune response due to lack of effective delivery of the vaccine candidate inside the cell. In this context, we evaluated a number of formulation and delivery devices and finally were able to eliminate this concern by using a needle-free injector system (NFIS) called Pharmajet Tropis® for the delivery of the vaccine.

Apart from effectively delivering the vaccine, the device also had certain other advantages like:

1. Eliminations of needle prick and making vaccinations painless.
2. Reducing the risk of disease transmission due to the use of contaminated needles.
3. Needle free delivery also contributes significantly to the reduction of sharp waste management.

As mentioned earlier, as there was no DNA vaccine approved for human use and there was no established platform available to us, we developed our product profile characteristic and embarked on this interesting product development journey.

10.2 Early Development

The designed and synthesized SARS-CoV-2 DNA vaccine candidate (ZyCoV-D) comprises a plasmid DNA vector carrying the gene encoding the spike protein (S) of the SARS-CoV-2 virus. The vector used for the development of ZyCoV-D was developed as per the Food and Drug Administration (FDA) document, "Considerations for Plasmid DNA Vaccines for Infectious Disease Indications" and was already demonstrated to be safe in several clinical trials, which gave the confidence about the safety of the vector backbone (Fig. 10.1).

The large-scale production of plasmid DNA was carried out using a prokaryotic expression system.

Buffers of different ionic strength and pH, stabilizers, pharmaceutical excipients such as bupivacaine hydrochloride, polysaccharides, polyols, adjuvants, and liposomal formulations were tried during the formulation development. The

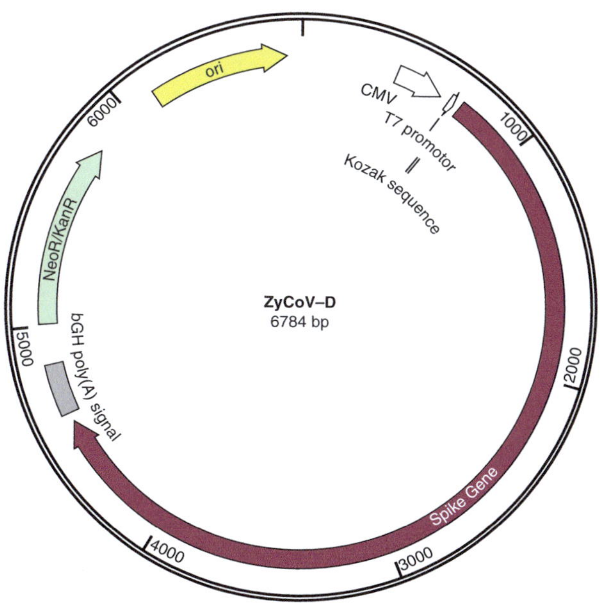

Fig. 10.1 Plasmid vector backbone of ZyCoV-D vaccine

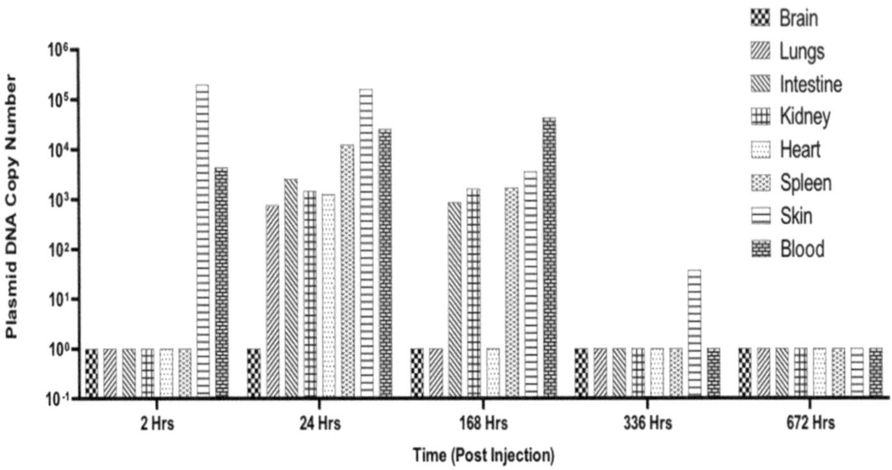

Fig. 10.2 Biodistribution study of ZyCoV-D vaccine in Wistar rats

stability and immunogenicity potential of various formulations were tested in mice, rats, guinea pigs, and rabbits before concluding on the final formulation.

The final vaccine candidate was evaluated in mice, guinea pigs, and rabbits and was found to induce neutralizing antibodies against SARS-COV-2 virus.

The vaccine also elicited a T-cell response complementary to antibody response in the mice model, as demonstrated by IFN-γ ELISPOT [1]. This finding was in line with the known fact that successful DNA vaccination induces both humoral and cellular responses and would be vital for providing long-term protection to the host post-vaccination.

Biodistribution studies were carried out for the vaccine candidate in Wistar rats using a single dose of 0.5 mg and 1.0 mg administered by intradermal route. The levels of plasmid DNA in various tissues and organs were measured at different time points using the validated real-time PCR method. Post intradermal injection, the plasmid was found to be absent from most of the organs within 2 weeks except the site of injection, which also cleared off by 4 weeks post-injection [1] (Fig. 10.2).

Our findings were in line with other plasmid DNA vaccine candidates including HIV-1, Ebola, Severe Acute Respiratory Syndrome (SARS), and a West Nile Virus candidate developed as reported earlier [10]. The study suggested that the plasmid DNA did not integrate in the host and was very safe besides being immunogenic.

10.3 Process Development

Upstream process development and optimization was initiated at a laboratory-scale in a shake flask, followed by 3 L fermentation scale. Subsequently, the process was scaled up to 15 L and then 150 L fermentation scale for the initial clinical developmental studies. Finally, the commercial-scale production was established at 1000 L and 5000 L fermentation scale.

Meanwhile, for downstream process development, a simplified, scalable process was developed involving cell lysis, removal of impurities like host cell proteins, endotoxins, RNA, etc. using various lysis and neutralizing buffers, chromatographies, and tangential flow filtration steps. The plasmid DNA was subsequently concentrated in a suitable buffer system to maintain a high concentration and optimal solution viscosity. Finally, the antigen was sterile-filtered to provide the final bulk, which was tested as per the quality control specifications laid out to comply with the available guidelines (Fig. 10.3).

The sterile filtered bulk was subsequently converted into the final finished product (drug product/ vaccine) as per the requirements and tested as per the available guidelines.

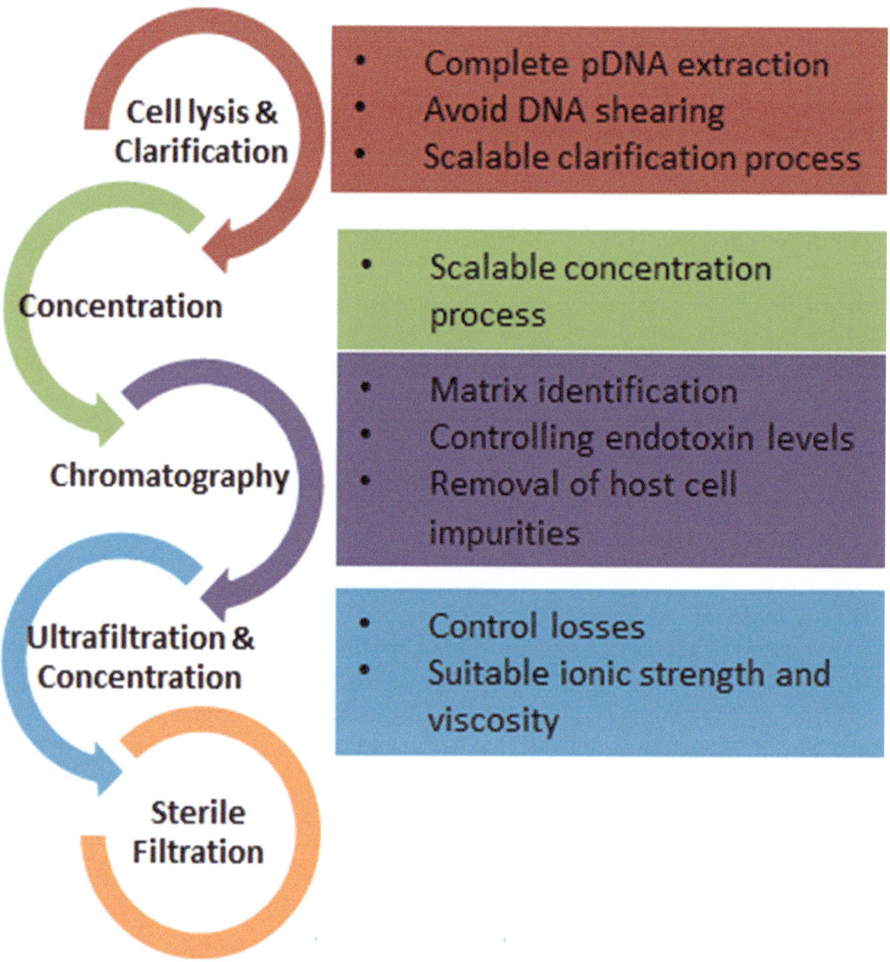

Fig. 10.3 Purification process flow and process key parameters

Fig. 10.4 ZyCoV-D drug substance vaccine manufacturing facility

10.4 Manufacturing Facility

A state-of-the art, fully automated drug substance manufacturing plant was set up in Ahmedabad. It is one of the most complex projects designed and executed during the pandemic with in-house capabilities. The plant has two 5 KL fermentation capacity and houses up to 30 KL capacity process vessels. It also houses one of the largest cross-flow filtration systems and the largest column chromatography systems in the country (Fig. 10.4). Furthermore, to reduce the dependency on import, we collaborated with the best of technology companies in the world to set up assembly and packaging lines to manufacture PharmaJet Tropis® device-disposable components in Baroda, Gujarat.

10.5 Preclinical Toxicological Studies

The preclinical toxicological studies were done to demonstrate the safety and tolerability of the ZyCoV-D vaccine. The outline of the study design was reviewed and approved by the Review Committee on Genetic Manipulation (RCGM), Department of Biotechnology, Ministry of Science & Technology, Government of India. The study was performed in OECD-GLP-certified and AAALAC-accredited facilities at Zydus Research Centre, Ahmedabad, India.

Repeat dose toxicity study of ZyCoV-D vaccine was conducted in Wistar rats and New Zealand white rabbits by intramuscular and intradermal routes. During the study, the animals were immunized four times by intramuscular or intradermal route on days 0, 9, 18, and 27 with different doses (maximum 1.5X and 3.0X of human dose in rats and rabbits, respectively).

Animals were observed for mortality and morbidity during the experimental period. Parameters evaluated included in-life observations such as clinical signs, detailed clinical examination, ophthalmic examination, body temperature, local tolerance effects, body weight, feed consumption, and neurobehavioral observations. Clinical pathology investigations were carried out at 48 h after the first dose and at the end of the treatment (after four doses). Terminal activities included organ weight measurements and detailed gross and histopathological examination. Immunogenicity assessment was performed both pre- and post-treatment periods.

Repeat intramuscular and intradermal injections of ZyCoV-D vaccine appeared to be safe and well tolerated in both rats and rabbits without any apparent signs of systemic toxicity, i.e., no serious adverse reactions or major concerns including gross and histopathological evaluations. Moreover, all animals elicited immunogenic response with antigen-specific antibodies in sera samples against the SARS-CoV-2 spike protein.

10.6 Non-Human Primate Challenge Studies

Rhesus macaques were immunized intradermally with different dose strengths (1 mg and 2 mg) of ZyCoV-D vaccine or placebo by the needle-free injection system (NFIS) and needle injection system (NIS) on days 0, 28, and 56. Subsequently, the animals were challenged with live SARS-CoV-2 virus after 15 weeks of the first immunization. Animals were monitored for morbidity/mortality once daily, and body weight was also recorded. Blood, nasal swab, throat swab, and bronchoalveolar lavage fluid specimens were collected on days 0, 1, 3, 5, and 7, post-challenge to determine the immune response and viral clearance. The 2 mg dose (NFIS) group elicited significant titers of IgG and neutralizing antibodies after immunization with enhancement in titers post-virus challenge. Besides this, it also induced increased lymphocyte proliferation and cytokine response. The significant reduction in viral load was observed post-SARS-CoV-2 challenge. The study demonstrated that a 2 mg dose of vaccine administration by NFIS was immunogenic and had prophylactic potential [11].

10.7 Clinical Trials

Once preclinical studies demonstrated that our DNA vaccine induces antibody response including neutralizing antibodies (NAB) against SARS-CoV-2 and also provides Th-1 response as evidenced by elevated IFN-g levels, we initiated clinical development of the vaccine, which was fast-tracked using an adaptive trial design with the combined Phase 1/2 protocol.

10.7.1 2 mg Three-Dose Clinical Trials

Phase 1 trial was conducted at Zydus Research Centre, Ahmedabad, India, after getting permission from CDSCO. The primary objective of the study was to evaluate the safety and tolerability of the ZyCoV-D vaccine candidate. The secondary objective of the study was to evaluate the seroconversion rate using IgG antibodies against spike S1 antigen, NAB titers, and IFN-g cellular immune responses after three doses of the vaccine. Eligible participants were healthy adults aged between 18 and 55 years; body weight > 50 kg; body mass index of between 18.5 and 29.9 kg/m^2. Participants were allocated to treatment arms sequentially in a nonrandomized, open-label, dose escalation manner (arm 1 [1 mg: Needle]; arm 2 [1 mg; Needle-Free Injection System {NFIS}]; arm 3 [2 mg; Needle], and arm 4 [2 mg; NFIS]). Participants were immunized intradermally with three doses of vaccine 4 weeks apart on days 0, 28, and 56. Follow-up visits were scheduled after each vaccination until day 84 (end-of-study). A total of 48 subjects were enrolled. There were no deaths or SAEs reported in Phase 1/2 of the study. The number of subjects with at least one solicited adverse event and at least one unsolicited adverse event was 7 (14.58%) and 6 (12.5%), respectively.

The proportion of subjects who seroconverted based on IgG titers on day 84 (i.e., 28 days after third vaccine doses) was 4 (36.36%), 4 (33.33%), 10 (100.00%), and 8 (80.00%) in the treatment Arm 1 (1 mg: Needle), Arm 2 (1 mg: NFIS), Arm 3 (2 mg: Needle), and Arm 4 (2 mg: NFIS), respectively.

This suggested a higher seroconversion rate with the 2 mg vaccine dose, irrespective of the method of delivery, compared to the 1 mg vaccine dose. In this study, ZyCoV-D vaccine, when administered intradermally via NFIS at 2 mg dose, showed peak cellular response in terms of IFN-g ELISPOT assay on day 56 with 41.5 spot-forming cells (SFC) per million PBMCs and was maintained till day 84 with a median 45.5 SFC per million PBMC [8].

The Phase 2 trial was initiated based on the data from the Phase 1 study and recommendation from the Data Safety Monitoring Board. The Phase 2 study was a four-arm, placebo-controlled, and assessor blind study conducted at eight study centers across India on OPD basis. Healthy subjects aged ≥12 years who were negative against SARS-CoV2 antibodies were enrolled. All the subjects were randomized in 4:1 ratio to receive either ZyCoV-D vaccine or matching placebo in one of the following four treatment arms: Arm 1 (1 mg: Needle), Arm 2 (1 mg: NFIS), Arm 3 (2 mg: Needle), and Arm 4 (2 mg: NFIS). The primary objective of this study was to evaluate the immunogenicity of the ZyCoV-D vaccine candidate. The secondary objective was to evaluate the safety and neutralization antibody response of the ZyCoV-D vaccine candidate. Majority of adverse events were mild or moderate in intensity and resolved. Overall, there was no meaningful difference observed in the incidence and severity of AEs between the treatment arms. The proportion of seroconverted subjects based on IgG titers on day 84 (i.e., 28 days after third vaccine doses) was 122 (72.62%), 138 (82.63%), 137 (82.64%), and 157 (91.28%) in the treatment Arm 1 (1 mg: Needle), Arm 2 (1 mg: NFIS), Arm 3 (2 mg: Needle), and Arm 4 (2 mg: NFIS), respectively. The proportion of seroconverted subjects based on NAB on day 84 (i.e., 28 days after third vaccine doses) was 7 (31.82%), 10 (45.45%), 11 (55.00%), and 16 (88.89%) in the treatment Arm 1 (1 mg: Needle), Arm 2 (1 mg: NFIS), Arm 3 (2 mg: Needle), and Arm 4 (2 mg: NFIS), respectively. The proportion of subjects seroconverted based on IgG titers on day 224 (i.e., 168 days after third vaccine doses) was 103 (65.19%), 108 (68.79%), 114 (71.70%), and 144 (88.34%) in the treatment Arm 1 (1 mg: Needle), Arm 2 (1 mg: NFIS), Arm 3 (2 mg: Needle), and Arm 4 (2 mg: NFIS), respectively. The seroconversion rate of IgG in 2 mg NFIS arm on day 56, day 70, day 84, day 140, and day 224 was higher than that in the treatment arms 1, 2, and 3.

After reviewing the data of Phase 1 and 2 studies, a Phase 3 multicentric, double-blind, randomized, placebo-controlled trial of ZyCoV-D was conducted at 49 centers in India. Healthy participants aged 12 years and above were enrolled. A total of 27,703 participants were enrolled and randomly assigned (1:1) to receive either ZyCoV-D vaccine (2 mg per dose) or placebo. All demographic characteristics were comparable between the treatment groups. The primary outcome was the number of participants with the first occurrence of symptomatic RT-PCR-positive COVID-19 28 days after the third dose, until the targeted number of cases (interim analysis $n = 79$, full analysis $n = 158$) have been achieved. Secondary endpoints were first occurrences of asymptomatic COVID-19 cases, severe COVID-19 cases, moderate COVID-19 cases, mild COVID-19 cases, and virologically confirmed COVID-19 deaths 28 days after the third dose of vaccine until the targeted number of cases had been achieved. Immunogenicity assessment included the seroconversion rate based on IgG against S1 antigen by ELISA, geometric mean titer (GMT), geometric mean fold rise (GMFR), and neutralizing antibody titers, as described previously. Safety assessment included the incidence and severity of solicited and unsolicited adverse events after each dose and the incidence of serious adverse events throughout the study.

All enrolled participants received three doses of ZyCoV-D vaccine (2 mg per dose) or placebo intradermally via a needle-free injection system 28 days apart (days 0, 28, and 56). The ZyCoV-D vaccine efficacy was found to be 66.6% (95% CI 47.6–80.7).

The proportion of participants who achieved seroconversion at day 84 was higher in the ZyCoV-D group ($n = 126$ [93.33%]) when compared with the placebo group ($n = 68$ [52.31%]). The antibody concentration defined by GMT based on IgG at day 84 was higher in the ZyCoV-D group (952.67 EU, 95% CI 707.94–1282.00) than the placebo group (154.82 EU, 91.25–262.70). The increase in antibody titer, as defined by GMFR, at day 84 was higher in the ZyCoV-D group (136.10, 95% CI 101.13–183.14]) than the placebo group (22.12, 13.04–37.5).

The immunogenicity response at day 84 in the group aged 12–17 years was higher than the overall participant population (IgG seroconversion 100% vs 93.33%, respectively; GMT 2083 EU vs 952.67 EU, respectively; and GMFR 297.65 vs 136.10, respectively).

The occurrence of solicited adverse events was similar between the treatment groups (623 [4.49%] in the ZyCoV-D group vs 620 [4.47%] in the placebo group). The occurrence of unsolicited adverse events was also similar between the treatment groups (454 [3.27%] in the ZyCoV-D group vs 463 [3.34%] in the placebo group). The majority of adverse events were mild to moderate in intensity in the ZyCoV-D group and the placebo group. No difference was observed with respect to successive dosing within each group or between the treatment groups.

The safety profile in terms of solicited and unsolicited adverse events in adolescent participants (12–17 years), participants older than 60 years, and participants with comorbidities was similar to the general study population. The most commonly reported local solicited adverse events were pain at the injection site, redness, swelling, and itching. The most commonly reported solicited systemic adverse events were headache, fever, muscle pain, and fatigue.

The Phase 3 trial showed that the three-dose regimen of ZyCoV-D was efficacious, safe, and immunogenic. The Phase 3 study provides the first evidence that the induction of immune responses against the spike protein using a plasmid construct carrying the spike-S-gene provides protection against COVID-19 in humans. ZyCoV-D could significantly contribute to the efforts to contain the COVID-19 outbreak in India and globally [3].

10.7.2 3 mg Two-Dose Clinical Trials

With an aim to reduce in-clinic visits for vaccination, a new two-dose regimen (3 mg per dose) with the dose administration at day 0 and day 28 was evaluated in subsequently conducted Phase I/II and Phase III clinical trials. The total dose administration in both the dosing regimens remains the same, i.e., 6 mg administered intradermally.

In the Phase I/II clinical trial conducted with the two-dose regimen (3 mg per dose), 150 subjects were enrolled, of which 100 subjects were administered ZyCoV-D vaccine, while the remaining 50 subjects were administered placebo. The age of subjects enrolled in the ZyCoV-D group ranged from 18 to 58 years while that in the placebo group ranged from 18 to 59 years. The male/female distribution in the two groups was similar. All the subjects enrolled in the study were healthy and did not have clinically significant disease at the time of vaccination.

The primary outcome was the seroconversion rate based on of IgG antibodies against the S1 antigen (by ELISA) at day 56 (28 days after second vaccine administration) which was significantly higher in the ZyCoV-D group as compared to the placebo group. The seroconversion rate in the ZyCoV-D group was 91.2% in those who were seronegative at baseline, while it was 87.6% in all subjects taken together.

The seroconversion rate based on neutralizing antibodies at day 42 and day 56 was also evaluated as a secondary outcome. The seroconversion rate based on neutralizing antibodies was 80.0% and 87.1% at day 42 and day 56, respectively, in the subjects who were seronegative at the baseline. The 3 mg–two-dose regimen also produced a higher level of antibodies in terms of GMTs at all time points post-vaccination.

A total of 48 adverse events were reported in 16 subjects in the ZyCoV-D group (adverse event rate, 16.0%), and 36 adverse events were reported in 7 subjects in the placebo group (adverse event rate, 14.0%) during the 56-day study period. All the adverse events reported in the study in both the ZyCoV-D and the placebo group were "mild" in severity. All the reported adverse events resolved completely with/

without symptomatic treatment during the study period. Overall, the safety profile of ZyCoV-D vaccine was comparable to placebo in this Phase I/II study.

Based on the immunogenicity and safety results in the above Phase I/II clinical trial, a further Phase III clinical trial was conducted to establish the immunogenicity and safety of the new two-dose regimen (3 mg per dose) in a larger set of population. In the Phase III study, a total of 3000 subjects were enrolled, among which 1193 were of the ≥12 to <18 years age group. Among these 3000 subjects, 150 subjects were considered for immunogenicity analysis (75 adults and 75 subjects ≥12 to <18 years of age).

The primary outcome was to establish non-inferiority based on the seroconversion rate based on IgG antibodies against S1 antigen (by ELISA) and neutralizing antibodies at day 56 of the 3 mg–two-dose regimen against the day 84 data from the Phase III clinical trial conducted with the 2 mg–three-dose regimen (ongoing at that time). The seroconversion rate based on IgG antibodies against S1 antigen (by ELISA) was 95.3% at day 56 of the 3 mg–two-dose regimen and 93.3% at day 84 with the 2 mg–three-dose regimen, while the seroconversion rate based on neutralizing antibodies was 91.6% at day 56 of the 3 mg–two-dose regimen and 88.0% at day 84 with the 2 mg–three-dose regimen. The seroconversion rate based on neutralizing antibody titers at day 56 in subjects seronegative at baseline was 91.5% in the total population and was also found to be similar in adult and pediatric cohorts separately. The cumulative seroconversion rate based on neutralizing antibodies including both seronegative and seropositive subjects at baseline was found to be 89.3% in the total population, 86.7% in adult cohort, and 92.0% in pediatric cohort. The GMTs based on the neutralizing antibodies at day 56 in subjects seronegative at baseline were 136.0 in the total population. The GMT based on neutralizing antibodies was found to be higher in the pediatric cohort as compared to that in the adult cohort.

Subsequently, the statistical analysis confirmed non-inferiority of the 3 mg–two-dose regimen versus the 2 mg–three-dose regimen.

All the adverse events recorded till day 56 after vaccination have been considered for safety evaluation. From a safety point of view, a total of 339 adverse events were reported in 245 subjects (adverse event rate, 8.2%). All the adverse events reported in the study were "Mild" in severity. All the reported adverse events resolved completely with/without symptomatic treatment during the study period. On tolerability front also, the 3 mg–two-dose regimen was similar to the 2 mg–three-dose regimen.

10.7.3 Single-Dose Heterologous Booster Clinical Trials

Following the recommendation of booster dose vaccination against COVID-19 by the Government of India, a further Phase III clinical trial was conducted to evaluate the immunogenicity and safety of a single 3 mg dose of ZyCoV-D administered as a heterologous booster dose in subjects who have already received two doses of

Covaxin/Covishield and compare that with the homologous booster dose of Covaxin/Covishield.

A total of 576 subjects were enrolled in the study. Among these, 288 subjects each were enrolled in the Covaxin cohort and the Covishield cohort. Among the 288 subjects in the Covaxin cohort, 144 subjects were randomized in the ZyCoV-D (heterologous booster) arm, while the remaining 144 subjects were randomized in the Covaxin (homologous booster) arm. Likewise, among the 288 subjects in the Covishield cohort, 144 subjects were randomized in the ZyCoV-D (heterologous booster) arm, while the remaining 144 subjects were randomized in the Covishield (homologous booster) arm.

The primary outcome was to establish noninferiority of the seroconversion rate (based on IgG by ELISA and neutralizing antibodies) at 28 days after the administration of the single 3 mg dose of ZyCoV-D administered as a heterologous booster as compared to the homologous booster of Covaxin / Covishield in subjects who had received two doses of Covaxin / Covishield for primary vaccination, respectively. The neutralizing antibodies were assessed against both Wuhan and Omicron virus strains.

The seroconversion rates were in the range of 49% to 65% in the ZyCoV-D arm, while it was in the range of 43% to 50% in the Covaxin arm of the Covaxin cohort. Likewise, seroconversion rates were in the range of 49% to 63% in the ZyCoV-D arm, while it was in the range of 42% to 54% in the Covishield arm of the Covishield cohort. Based on the predefined non-inferiority margin of −10%, heterologous booster vaccination with a single dose of 3 mg ZyCoV-D was found to be non-inferior to the homologous booster vaccination with Covaxin or Covishield. The heterologous booster vaccination with ZyCoV-D was also found to be non-inferior to the homologous booster vaccination with Covaxin or Covishield based on the GMTs of IgG by ELISA and neutralizing antibodies at 28 days after booster vaccination based on the predefined non-inferiority margin of 0.67 for GMT ratios.

All the adverse events recorded till day 84 after vaccination were considered for safety evaluation. A total of seven adverse events were reported in seven subjects in the ZyCoV-D arm of the Covaxin group, and 10 adverse events were reported in 10 subjects in the Covaxin arm of the Covaxin group. However, in the Covishield group, 14 adverse events were reported in 12 subjects in the ZyCoV-D arm, and 20 adverse events were reported in 18 subjects in the Covishield arm. Most of the adverse events reported in the study till date were "Mild" in severity. All the reported adverse events resolved completely with/without symptomatic treatment during the study period.

10.8 Licensure

The largest Phase III clinical trial of any COVID-19 vaccine was conducted for ZyCoV-D in India, and it resulted in an Emergency Use Authorization (EUA) of the vaccine for ≥12 years population as a three-dose regimen in August 2021. Subsequently, a two-dose regimen was also evaluated in Phase III clinical trials, and the vaccine was found to be non-inferior to a three-dose regimen in terms of safety and immunogenicity which led to the vaccine receiving a EUA for ≥12 years population as a two-dose regimen as well in April 2022.

Recently, we have also received a EUA for ZyCoV-D vaccine as a heterologous booster to adults vaccinated with Covaxin or Covishield based on the promising clinical trial results.

10.9 Way Forward

The global vaccine research and development (R&D) during the COVID-19 pandemic was unprecedented in scale and speed. Due to urgent requirements, vaccine developers and regulators did fundamental changes in their approach toward vaccine development, production, and approval process, which resulted in developing of number of COVID-19 vaccines in a very short span of 12 to 18 months.

This also led to the emergence of new vaccine platforms like viral vectored vaccines, nucleic acid vaccines, chimeric and combination vaccines besides new adjuvants, formulations, and delivery devices. Indeed, these developments have revolutionized the field, and more advancements are envisaged in these areas in the coming years.

As far as the DNA vaccine platform is concerned, though over the years, lot of advancements had happened in the field, but COVID-19 pandemic accelerated this field of science multifold. Advancements in the plasmid DNA design, construction, manufacturing, and delivery not only improved the immunogenic response but also highlighted the importance of plug and play technology and thermostability of DNA vaccines, an important hallmark for the development of vaccines for masses.

The licensure of World's first human DNA vaccine, ZyCoV-D, marks an inflection point in this field as it has addressed most of the major drawbacks encountered by researchers in this field earlier [4, 5, 9]. This has also created history and has put our country on the World map to be the first to develop a human DNA vaccine.

The success of our DNA platform technology has ignited hope for vaccines against existing diseases like cancer, HIV, Hepatitis-C, and tuberculosis where no or limited vaccines are available.

Furthermore, the development and advancement of DNA vaccine platform has also now opened the door for future preparedness in pandemic situation as climate change and emergence of new zoonotic diseases are considered a major threat to mankind.

Acknowledgments We express our gratitude to Niti Aayog, National Technical Advisory Group on Immunization (NTAGI), CDSCO, and Department of Biotechnology for continuous guidance and support.

We also thank National Biopharma Mission, BIRAC, Department of Biotechnology, Government of India, for providing grant for preclinical, Phase I, and Phase III clinical trials.

We also thank Interactive Research School for Health Affairs (IRSHA), Pune, for testing of clinical samples for neutralization assays and ICMR-National Institute of Virology, Pune, for the conduct of NHP Challenge study.

We thank our partner Pharma Jet Inc. Golden, CO, USA for providing us the Pharma Jet Tropis® device.

Competing Interest The authors have no conflicts of interest to declare that are relevant to the content of this chapter.

References

1. Dey A, Chozhavel Rajanathan TM, Chandra H, et al. Immunogenic potential of DNA vaccine candidate, ZyCoV-D against SARS-CoV-2 in animal models. Vaccine. 2021;39(30):4108–16.
2. Flingai S, Czerwonko M, Goodman J, Kudchodkar SB, Muthumani K, Weiner DB. Synthetic DNA vaccines: improved vaccine potency by electroporation and co-delivered genetic adjuvants. Front Immunol. 2013;4:354.
3. Khobragade A, Bhate S, Ramaiah V, et al. Efficacy, safety, and immunogenicity of the DNA SARS-CoV-2 vaccine (ZyCoV-D): the interim efficacy results of a phase 3, randomised, double-blind, placebo-controlled study in India. Lancet. 2022;399(10332):1313–21. https://doi.org/10.1016/S0140-6736(22)00151-9.
4. Ledesma-Feliciano C, Chapman R, Hooper JW, et al. Improved DNA vaccine delivery with needle-free injection systems. Vaccines (Basel). 2023;11(2):280.
5. Liu MA. DNA vaccines: a review. J Intern Med. 2003;253(4):402–10.
6. Martin JE, Louder MK, Holman LA, et al. A SARS DNA vaccine induces neutralizing antibody and cellular immune responses in healthy adults in a phase I clinical trial. Vaccine. 2008;26(50):6338–43.
7. Modjarrad K, Roberts CC, Mills KT, et al. Safety and immunogenicity of an anti-Middle East respiratory syndrome coronavirus DNA vaccine: a phase 1, open-label, single-arm, dose-escalation trial. Lancet Infect Dis. 2019;19(9):1013–22.
8. Momin T, Kansagra K, Patel H, et al. Safety and immunogenicity of a DNA SARS-CoV-2 vaccine (ZyCoV-D): results of an open-label, non-randomized phase I part of phase I/II clinical study by intradermal route in healthy subjects in India. EClinicalMedicine. 2021;38:101020. https://doi.org/10.1016/j.eclinm.2021.101020.
9. Shedlock DJ, Weiner DB. DNA vaccination: antigen presentation and the induction of immunity. J Leukoc Biol. 2000;68(6):793–806.
10. Sheets RL, Stein J, Manetz TS, et al. Biodistribution of DNA plasmid vaccines against HIV-1, Ebola, Severe Acute Respiratory Syndrome, or West Nile virus is similar, without integration, despite differing plasmid backbones or gene inserts. Toxicol Sci. 2006;91(2):610–9.
11. Yadav PD, Kumar S, Agarwal K, et al. Needle-free injection system delivery of ZyCoV-D DNA vaccine demonstrated improved immunogenicity and protective efficacy in rhesus macaques against SARS-CoV-2. J Med Virol. 2023;95(2):e28484.

11 Make in India Vaccines: A True Story of Atmanirbhar Bharat—MADE IN INDIA, MADE FOR INDIA: Establishment of the Nation's First Pandemic-Ready mRNA-Based Platform Technology

Sanjay Singh and Tathagata Mukherjee

11.1 Company Overview

Gennova Biopharmaceuticals Ltd., headquartered in Pune, India, is a biotechnology company dedicated to developing, producing, and commercializing biotherapeutics to address life-threatening diseases across various indications.

Gennova is transforming healthcare by creating efficient and effective solutions for manufacturing and successfully commercializing biotherapeutics across cardiovascular, neurology, nephrology, and oncology segments. Incorporating recombinant DNA technologies and innovative biomanufacturing practices, the company manufactures its recombinant products using bacterial, yeast, and mammalian-based expression platforms and has developed deep expertise in perfusion-based continuous manufacturing technologies. Gennova has commercialized nine products, including biosimilars and pioneering products.

The state-of-the-art facilities, R & D capabilities, and knowledge base development have empowered Gennova to take a product from bench to bedside/gene to market. Gennova has a team of experienced and capable scientists and technocrats who have developed capabilities across scientific, manufacturing, regulatory, clinical, and business arenas that have made it successful in the Indian market.

Gennova's socio-economic responsibility has propelled it to establish the Global Health Initiative program, under which Gennova has established itself as a preferred research and cGMP manufacturing partner in the area of vaccine development to address some of the world's most challenging diseases like malaria, leishmania, flu, etc.

Gennova has developed two mRNA vaccines, GEMCOVAC®-OM and GEMCOVAC®-19, approved for Restricted Use in Emergency Situations. They are India's first mRNA vaccines and the world's first thermostable mRNA vaccines

S. Singh (✉) · T. Mukherjee
Gennova Biopharmaceuticals Limited, Pune, India
e-mail: Sanjay.Singh@gennova.co.in

© The Editor(s) (if applicable) and The Author(s), under exclusive license to Springer Nature Singapore Pte Ltd. 2025
R. Swarup et al. (eds.), *At War with the Single Strand*,
https://doi.org/10.1007/978-981-96-8122-8_11

against COVID-19. Further, this technology offers a rapid development path that will empower Gennova to combat future pandemic outbreaks.

Gennova aspires to transform millions of people's healthcare, primarily through technological solutions, employing AI-assisted innovation in biomanufacturing, diagnosis, and treatment for a better outcome (https://gennova.bio/).

It was a foggy evening in November 2020. The time was 7:00 pm. We sat in the office immersed in myriads of discussions related to mRNA vaccines. Suddenly, we heard a loud cheer from the conference room at our Vaccine Development Research Centre in Pune. The results from the immunogenicity assays from the plasma samples of the mice experiments done with our prototype mRNA-based vaccine were due. We were eagerly awaiting these results as this would be the first time in our hands that the proof of concept would be established.

There was a knock on the door, and without waiting for a reply, the entire scientific leadership team came barging into the room. They were all immensely excited and happy. They all tried to speak simultaneously to convey that the immunogenicity results were very, very positive. We got not only good titers but also good neutralizing results, which effectively meant that our vaccine was immunogenic.

One of them pumped his fist in the air and said—we made it! We made India's first mRNA vaccine!

Indeed, it was a moment to celebrate. Throughout the COVID-19 pandemic, even in its early days when the risks were severe, our team worked day in and day out, fearlessly, to develop India's first mRNA vaccine. They deserved to celebrate. They had poured their blood and sweat into this program, as did the other frontline workers, toward developing a pandemic-ready technology with promises of rapid development timelines.

Only if India had access to such a rapid development platform technology, the outcome of the effects of the pandemic in a population-dense country like India would have been different. But it is better late than never. Gennova embarked on the journey to develop this technology platform that could be utilized to develop a vaccine against viruses with a quick turnover time. The entire Gennova stood steadfast throughout this journey, and today, they have tasted the success of their efforts; we did have an mRNA vaccine that works.

By now, the news of Moderna and BioNTech-Pfizer changing the course of their mRNA technology for COVID-19 was out in the open. Little did the world know that Gennova had already embarked on a similar journey for using mRNA technology, a two-decade-old platform. We will only be the third in the world to come out with an mRNA-based vaccine for COVID-19.

So, the right to celebrate was manyfold during the crisis.

However, we still did not have an mRNA vaccine for India. When we took up the project, we wanted to make a vaccine for our fellow citizens. The truth was that we were far away from an mRNA vaccine for Indians. We had just begun a journey to a destination no one had traveled before.

Moderna's and BioNTech-Pfizer's mRNA-based COVID-19 vaccine required ultra-low temperature conditions for storage and distribution. India (including other LMICs) did not possess the vast and penetrating infrastructure required for deploying the mRNA-based vaccine. What is the use of having a vaccine that can only be given in metropolitan and highly urbanized cities? It was impossible to fathom deploying an mRNA-based COVID-19 vaccine that was contingent upon having access to ultra-low temperature storage conditions in the vaccination centers Pan-India.

mRNA is fragile and breaks down quickly, so vaccines based on this platform need to be stored at extremely low temperatures to assure efficacy. True, we were successful in making an mRNA-based vaccine that can be given in the arm, but ours were no better than BioNTech-Pfizer's and Moderna's, or for that matter CureVac, Imperial College London and few others who had now started their mRNA vaccine programs. We can certainly take this vaccine forward and test it in humans and roll it out, subject to regulatory approvals, but what will we have accomplished? We have made the third mRNA-based vaccine for COVID-19 that cannot be given to the masses—since it still has the conditional requirement of ultra-low temperature infrastructure.

We can only consider being successful in truly rising to the call of duty and serving our nation if we can make an mRNA vaccine formulation that is stable at 2–8 °C and can be distributed through the already existing refrigeration supply chain Pan-India (and in LMICs). This is the only way to effectively deliver mRNA-based vaccines to every corner of India, enabling even the most remote village in India to benefit from this technology.

When confronted with Hamlet's situation—"to be or not to be"—stating the facts in its most accurate form, no matter the consequence, is the most prudent thing to do.

Their success was acknowledged, and their efforts were praised. The team's excitement had ebbed a little as they realized that the expected reaction was not over the moon with the immunogenicity results.

They were told in plain and simple words—if Gennova's vaccine is not deployable to the citizens of India, then we do not want to take the vaccine to the market.

Even if we successfully make a vaccine that can be stored and distributed at 2–8 °C, we will not take it to the market if we do not successfully develop a pandemic-ready platform technology. Other vaccines were in the late stages of development, and we knew that we as a nation together would survive COVID-19, but the truth is that we were not prepared at the start of the pandemic to nip it in the bud. We as a nation need to develop and have ready access to a pandemic-ready technology that can be leveraged for future pandemics. Though the team's celebration was a little unreciprocated, there was a shine of pride and commitment in their eyes.

Like the rest of the world, the Indian economy has been hard hit by COVID-19. The impact of the COVID-19 pandemic on India has been largely disruptive in terms of health and economic productivity.

India's GDP shrank 7.3% in 2020–21, the worst performance of the Indian economy since independence [1]. The fact that India's growth rate in 2019 was among the highest makes the drop due to COVID-19 even more significant and noticeable [2].

The pandemic led to massive and reverse migration in different parts of the country, from urban to rural areas. Indian roads were flooded with laborers marching back to their villages, which enormously impacted the societal health, demography, and economy of rural India [3]. The demand and supply gap increased due to trade restrictions and labor mobility.

COVID-19 has caused much collateral damage with certain effects that may be uncovered in the future. Various hidden aspects include shortage of medicines and healthcare infrastructure for people suffering from other acute diseases, food insecurity, declining mental and physical health, effects on education, economic and industrial downfall, etc. [1].

So, the bottom line was that we needed to be prepared, for which we require a pandemic-ready technology. And we need to keep in mind that while developing a vaccine, affordability needs to be a high priority to make the vaccines globally accessible.

My team was asked to remember why we chose to develop the mRNA vaccine for COVID-19.

The obvious answers that came forth were that mRNA vaccines were one of the best technologies available, as

- mRNA vaccines are neutral to causing infection, i.e., immunogenic but not infective. So, the risk of getting an infection out of a vaccine jab is negligible.
- mRNA does not need to be integrated into the host cell genome to do its work; therefore, any possibility of host genome modification can be eliminated.
- mRNA is nonreplicative in nature; it will not cause any accumulation inside the cell.
- It decays metabolically by standard cellular mechanisms.

In addition to safety, they are highly efficacious because when injected into the muscle, the vaccine complex is taken up by the cells and bordering immune cells, and the mRNA is released inside the cell, where it is translated to a protein by the host machinery. The translated protein acts as an antigen, which is then recognized by the host immune system. Antigen-presenting cells, in coordination with helper T-cells, initiate a cascade of immunological events and lead to the formation of SARS-CoV-2-neutralizing antibodies as well as requisite durable cellular response.

One answer came out after some probing—mRNA vaccines are entirely synthetic, and therefore, they can be quickly manufactured inexpensively under cGMP conditions. This is important to ensure their "availability" and "accessibility" for mass vaccination on a sustainable basis, especially in developing nations like India.

The task for the team was now threefold:

(a) Developing an mRNA vaccine that is stable at 2–8 °C
(b) Developing a rapid pandemic-ready technology platform
(c) Developing independent unit operations that are easy to replicate and can be easily technology-transferred to others to manufacture the vaccine

These, we believe, were the only ways to democratize the vaccine for COVID-19 and all future pandemics that can potentially paralyze the world.

With zero dampening of their spirit, they rose in unison and said that they had an arduous task to accomplish and wanted to get going with it.

To cut a long story short, we developed the world's first thermostable mRNA-based COVID-19 vaccine. Gennova's mRNA vaccines are based on freeze-drying technology that enables storage at 2–8 °C, compared to other mRNA vaccines that would require ultra-low temperature conditions. Stabilizing the relatively large and fragile mRNA molecule on the nanoparticle was a daunting challenge. After investing hundreds of people-hours and applying expertise with innovative, cutting-edge science, Gennova succeeded in making the first-of-its-kind single-vial thermostable mRNA vaccine formulation within a year.

In addition to the thermostability property of the vaccine, the highlight of the mRNA vaccine is the self-amplifying nature, due to which it requires a low dose and has shown a robust immunogenic response to fight COVID-19 disease.

The mRNA vaccine manufacturing has four-unit operations that can be scaled independently (Fig. 11.1).

The first one is the production of the plasmid DNA encoding the antigenic protein, which will serve as the template to make the mRNA. Gennova has experience and access to microbial fermentation technology and developed an in-house process to produce plasmid DNA in substantial quantities. Once the DNA is purified, it can be stored at −20 °C. This is the only cell-based step in the whole manufacturing of mRNA. If plasmid DNA is considered a reagent sourced from outside, then the entire manufacturing of the mRNA is cell-free and synthetic.

The second step involves the production of mRNA by a well-established process called the in-vitro transcription (IVT) method, which requires five to seven

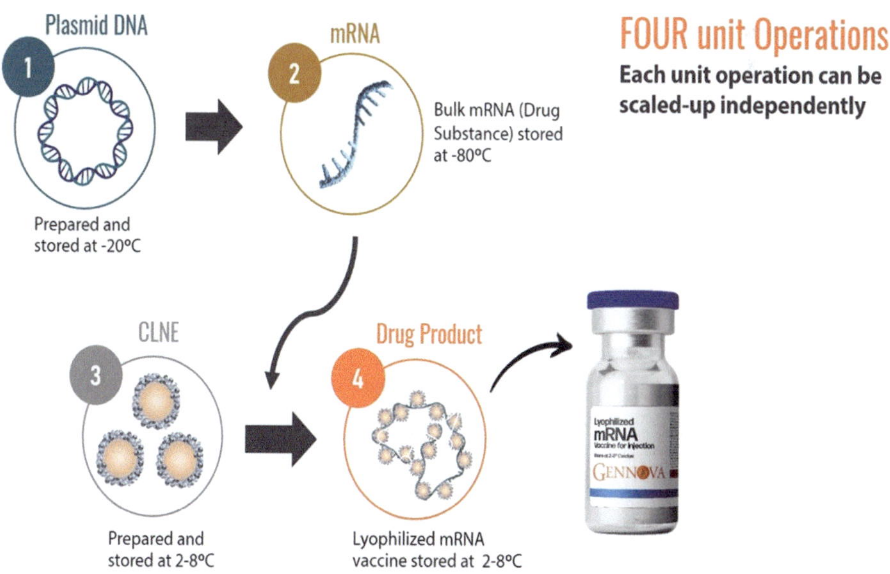

Fig. 11.1 Four-Unit Operation for mRNA vaccine manufacturing

enzymes (depending on the IVT process), small-molecule cofactors like SAM, GTPs, etc., and the building block of mRNA—the nucleotides (NTs)—as essential raw materials for production. This process was optimized for each step and ensured that Gennova's IVT process was readily scalable and economically viable. Once the mRNA is produced, it can be stored at −80 °C till further use for making the vaccine.

The third step involves the synthesis of the nano-lipid-based delivery system. Unlike other mRNA-based vaccines which use nano-lipid carriers (NLCs), which are essentially liposomal in nature and entrap the mRNA inside the micro-aqueous layers of their hydrophilic cores, Gennova has developed a lipid nano-emulsion that does not require entrapment of the mRNA. Because of the cationic nature of the lipid, Gennova's lipid emulsion is called Cationic Lipid Nano-Emulsion (CLNE). The mRNA encoding the viral "spike" protein is adsorbed on the surface of these nano-lipid emulsions, thereby easing manufacturability and minimizing losses. This reasonably scalable process was optimized and characterized in-house. The CLNE can be synthetically produced as a unit operation and can be stored at −20 °C till further use.

The last step involves combining the mRNA produced in step 2 with the CLNE produced in step 3 and then lyophilizing the vaccine to form a powder that can be stored at 2–8 °C. This can be reconstituted with sterile water for injection (WFI) before administration.

Each of these steps is readily scalable and amenable to technology transfer—a must for democratizing mRNA-based technology to eliminate the inequitable vaccine distribution globally.

It is worth mentioning that we were developing the process during the pandemic, and there was an acute shortage of raw materials for vaccine manufacturing due to the embargo imposed by the US and other nations. The mRNA technology was relatively new, and the world was not ready to cater to the global production requirement of raw materials. The supplies of critical materials were sparse, leading to high cost, foreign dependency, and long lead times—the limitation of ensuring the supply of raw materials for vaccine production needed to be circumvented.

Rising to the call of "Atmanirbhar Bharat" of the Honorable Prime Mister of India, Shree Narendra Modi, Gennova took steps to become independent of USA, China, and other countries for sourcing key raw materials for the production of the mRNA vaccine.

To ensure self-reliance, uninterrupted supply, and timely vaccine delivery, Gennova established a process for the development of IVT enzymes (guanylyltransferase, T7 RNA polymerase, DNase I, inorganic pyrophosphatase, RNase inhibitor, etc.) at the lab scale. We were poised for scale-up by the time the embargo was lifted due to the kind interference of the Government of India, and ready supply was obtained. Gennova also developed an in-house technology to produce small molecules (S-Adenosyl methionine, SAM, and rNTPs), as well as lipids like Dioleoyl-3-trimethylammonium propane (DOTAP), a key component for the nanoparticle carrier.

In developing the technology platform, Gennova's scientists successfully overcame the hurdles preventing rapid production and global distribution of the much-needed vaccine (Fig. 11.2). This is a must to eradicate any pandemic.

By the time all the hurdles mentioned above were circumvented and India's first mRNA vaccine against COVID-19, GEMCOVAC®-19, got approval for Restricted Use in Emergency Situations, the SARS-CoV-2 virus had mutated multiple times, giving rise to several variants. Successive waves of new infections and hospitalizations resulted from the emergence of highly infectious SARS-CoV-2 variants. These variants exhibit differences in infectivity, transmission, and the ability to evade the immune system. Mutations also affect the disease severity and, due to the ineffectiveness of vaccines that were developed against the ancestral strains, may pose serious public health concerns. GEMCOVAC®-19 was developed against the Wuhan strain and not against the emerging variants.

SARS-CoV-2 is still evolving, and the most recent variant, Omicron, not only spreads faster but is also the most antibody-resistant variant to date. This necessitated the development of an advanced vaccine effective against the new Omicron strain.

By 2022, the Indian population was immunized with vaccines produced against the ancestral strain. Moreover, the currently approved vaccines used as precautionary doses are against the ancestral strain. Although these will increase the antibody titers, their ability to neutralize the circulating Omicron variant of SARS-CoV-2 is limited.

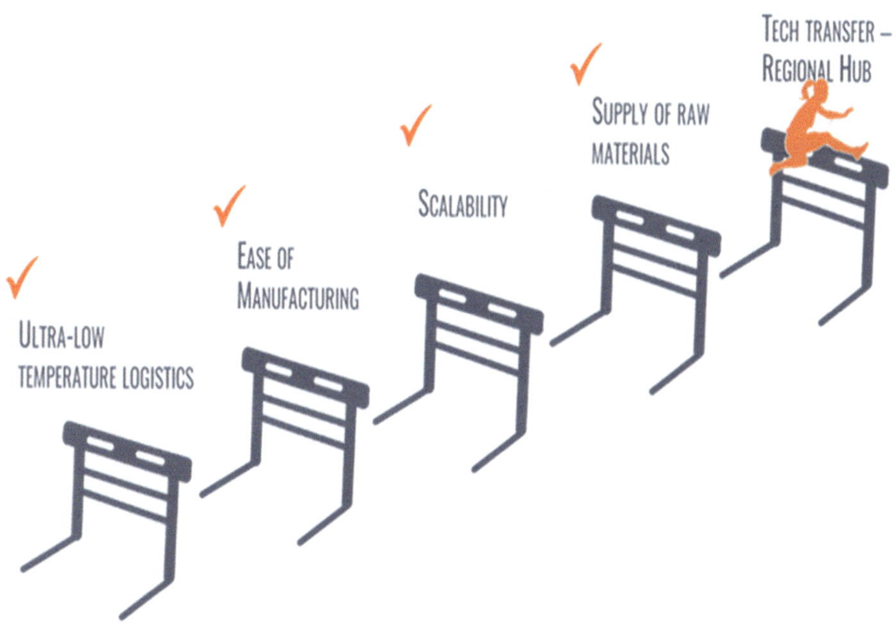

Fig. 11.2 Hurdles overcome while developing a Pandemic-Ready mRNA-based platform technology

India saw an increase of over 1000 new cases of COVID-19 every day, with several deaths recorded. As seen with other countries, there is a possibility of a new wave of infection. The availability of a booster dose to avert this wave is imperative.

A promising approach to tackle the current and future variants relies on a booster dose for long-lasting immunity.

The mRNA vaccine technology can be rapidly adapted to protect against emerging variants. It is highly suited to develop new variant-specific vaccines within a few months. The prevalence of multiple variants and their effect on the disease transmission dynamics and pathophysiology highlight the need for a booster, which is essential not only for high-risk groups but also for all eligible individuals.

Rising to the need, Gennova developed an Omicron-specific vaccine, GEMCOVAC®-OM, which can be administered as a booster in people above 18 years of age who have received two doses of COVAXIN® or COVISHIELD™ as their primary vaccination. Gennova's mRNA vaccine platform is modular and can rapidly develop vaccines against the virus's new variants.

GEMCOVAC®-OM was safe and well tolerated, with a significantly strong immune response against the virus. Gennova had successfully conducted a Phase II/III study to assess the safety and immunogenicity of GEMCOVAC®-OM developed for the SARS-CoV-2 Omicron variant in 3000 subjects in 20 centers across 13 cities when administered as a booster in participants who have received two doses of COVAXIN® or COVISHIELD™ as their primary vaccination.

Improving on the previous prototype vaccine, administered through intramuscular injection, GEMCOVAC®-OM is administered intradermally via a needle-free device called Tropis®, developed by PharmaJet. This device reduces the risk of needle injuries, cross-contamination, or needle phobia and has no disposal hazards. This improves the patient and caregiver's experience, allowing for mass vaccination in a very short time. Vaccines administered intradermally show better immunogenicity when compared to intramuscular administration.

We have created a rapid-development pandemic-ready mRNA-based platform technology to produce a "go-anywhere" vaccine stable at 2–8 °C. Gennova's established mRNA-based platform technology takes care of ultra-low temperature logistics, ease of manufacturing, scalability, and amenability to technology transfer. Using this platform technology, we have developed two mRNA vaccines against COVID-19, GEMCOVAC®-19, and an Omicron-specific booster, GEMCOVAC®-OM, approved for Restricted Use in Emergency Situations.

mRNA molecules are synthesized in vitro; therefore, the manufacturing cycle is shorter and relatively less expensive to ensure easy availability and accessibility for mass vaccination, especially in healthcare resource-constrained countries.

The mRNA platform can use templates for multiple antigens and overcome the limitations of using antigens from bacteria or viruses, as they may contain components that may elicit unwanted immune responses.

In addition to the obvious application to infectious diseases, mRNA-based vaccines have the potential for multiple administrations in patients with low intrinsic immunity, thereby opening bright prospects for developing novel therapeutics with the potential to treat or prevent diseases that are currently incurable.

The establishment of such a technology platform has not only empowered India to handle the COVID-19 pandemic but also ensured the preparedness for any future pandemic or endemic threat that may follow by utilizing its rapid development path. The speed of this platform technology has already been proven during the COVID-19 outbreak as Gennova successfully launched GEMCOVAC®-19 and GEMCOVAC®-OM as a booster dose within 2 years.

Creating an unprecedented "disease agnostic thermostable mRNA-based technology platform" empowers the world to prepare for future pandemics. The development of a versatile mRNA platform will prove hugely beneficial (Fig. 11.3).

Currently, mRNA vaccines are under investigation for prophylactic and therapeutic use. They can potentially be used not only for the prevention of infectious diseases but also for the treatment of various cancers and autoimmune diseases [4].

According to the clinical trials database at www.clinicaltrials.gov, there are over 400 clinical trials in various phases registered for the use of mRNA vaccines against different disease conditions. For COVID-19 alone, there are over 250 registered clinical trials for mRNA vaccines.

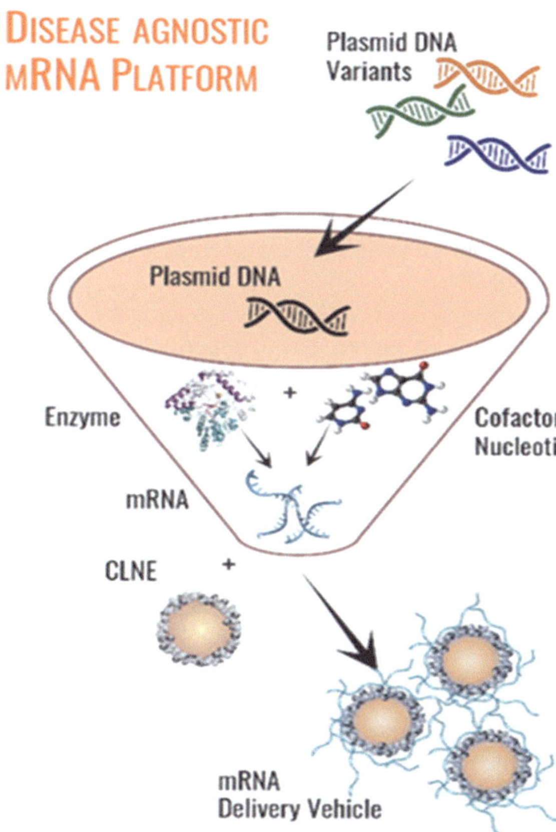

Fig. 11.3 Disease Agnostic mRNA technology

mRNA vaccines have marked the rise of a new era in vaccinology. Due to their ease of production, high potency, and safety, mRNA vaccines have emerged as a promising alternative to conventional vaccines.

Significant breakthroughs in using lipid nanoparticles to deliver mRNA molecules have allowed for improved stability, enhanced delivery to target cells, and increased potency of mRNA vaccines [5].

mRNA vaccine candidates against Influenza, Rabies, Zika, Malaria, Nipah, Tuberculosis, Chikungunya, and HIV are being tested in various phases of clinical trials. mRNA vaccines have successfully elicited B-cells capable of producing anti-HIV antibodies [6] and have been a significant breakthrough, indicating the vast potential of this technology. Currently, an mRNA vaccine for RSV (mRESVIA) has received FDA approval [7]. Thus, mRNA vaccines have offered hope for tackling some of the world's most pressing global challenges.

mRNA vaccines are also revolutionizing the field of cancer immunotherapy. The cancer mRNA vaccines encode tumor-specific antigens or neo-antigens unique to the patient's cancer cells. Once the mRNA vaccine is delivered to the patient's body, it is taken up by the antigen-presenting cells, which process the mRNA and present the encoded antigen on the surface. This helps T-cells to recognize and destroy the cancer cells [8]. Currently, several clinical trials are enrolling patients for various mRNA-based cancer vaccine therapy.

Scientists are also developing mRNA vaccines to treat autoimmune diseases like multiple sclerosis. Studies are also conducted to study the potential mRNA vaccine therapies in replacing missing proteins in the body in various metabolic and genetic disorders [9].

With the increasing prevalence of diseases, changing lifestyles, and increased population burden, the global mRNA vaccines and therapeutic market are expected to grow from USD 56.1 billion in 2022 to 66.2 billion in 2028, with a CAGR of 2.8% [10].

The emergence of mRNA vaccines has ushered in a new era of vaccine development, offering an unprecedented opportunity to address global health challenges. As we continue to explore the potential of this innovative technology, we can look forward to a future where mRNA vaccines play a pivotal role in shaping the healthcare landscape.

Our scientific efforts have been recognized and published in the journal *Nature Medicine* [11], where the outcome of the clinical trial that Gennova conducted to evaluate the safety and immunogenicity of GEMCOVAC®-OM as a booster in healthy adults has been highlighted. The results of this trial have demonstrated that GEMCOVAC®-OM is well-tolerated with no related serious adverse events and significantly boosts immune responses against the Omicron variant.

This indigenous platform technology developed by Gennova would not have been possible without the support of the Department of Biotechnology (DBT), Government of India, through the "Mission COVID Suraksha—the Indian COVID-19 Vaccine Development Mission" implemented by the Biotechnology Industry Research Assistance Council (BIRAC).

References

1. Impact TOI. https://timesofindia.indiatimes.com/readersblog/shreyansh-mangla/impact-of-covid-19-on-indian-economy-2-35042/.
2. India grt 2019. https://cep.lse.ac.uk/pubs/download/cp619.pdf.
3. Kumar R, Bharti N, Kumar S, Prakash G. Multidimensional impact of COVID-19 pandemic in India—challenges and future direction. J Fam Med Prim Care. 2020;9:5892–5.
4. Krienke C, et al. A noninflammatory mRNA vaccine for treatment of experimental autoimmune encephalomyelitis. Science. 2021;371:145–53.
5. Karam M, Daoud G. mRNA vaccines: past, present, future. Asian J Pharm Sci. 2022;17:491–522.
6. Venkatesan P. Preliminary phase 1 results from an HIV vaccine candidate trial. Lancet Microbe. 2021;2:e95.
7. mRNA RSV. https://www.fiercepharma.com/pharma/moderna-fda-nod-rsv-vaccine-grows-its-reach-beyond-pandemic#:~:text=Moderna's%20mRESVIA%20has%20secured%20FDA,late%20fall%20in%20the%20U.S.
8. Lorentzen CL, Haanen JB, Met Ö, Svane IM. Clinical advances and ongoing trials on mRNA vaccines for cancer treatment. Lancet Oncol. 2022;23:e450–8.
9. Qin S, et al. mRNA-based therapeutics: powerful and versatile tools to combat diseases. Signal Transduct Target Ther. 2022;7:166.
10. https://www.prnewswire.com/news-releases/mrna-vaccines-and-therapeutics-global-market-report-2023-sector-to-reach-66-2-billion-by-2028-at-a-2-8-cagr-301761281.html.
11. Saraf A, et al. An Omicron-specific, self-amplifying mRNA booster vaccine for COVID-19: a phase 2/3 randomised trial. Nat Med. 2024;30:1363–72.
12. https://gennova.bio/.

Make in India Vaccines: A True Story of Atmanirbhar Bharat—CORBEVAX™ Development and Deployment, A Highly Successful Partnership Between Biological E and BIRAC During the COVID19 Pandemic

12

Vikram Paradkar and Mahima Datla

Biological E Limited is an integrated vaccine development and manufacturing organization with nine WHO-prequalified vaccines. BioE manufactures all classes of vaccines and supplies more than two million doses of vaccines every day to more than 135 countries for deployment in childhood immunization programs.

Our work with Janssen Ad26.COV2.S COVID-19 vaccine:

Janssen, a division of Johnson & Johnson, initiated discussions with BioE during the early part of the COVID19 pandemic for becoming a manufacturing partner for vaccine production. BioE constructed a new manufacturing facility to produce the Adenovector Drug Substance (antigen) and modified two existing formulations and filling suites for the Drug Product manufacturing. During the initial phase of the pandemic, BioE faced major obstacles in the construction and qualifications of production facilities due to lockdowns, limitations in shipments, and restrictions for import of bioprocessing equipment. Still, we managed to complete both complex projects in time and initiated technology transfer for both Drug Substance and Drug Product. After the successful completion of technology transfer, comprehensive process validation was completed along with the extensive comparability of BioE-manufactured vaccine with that produced at the Janssen originator site. Both the manufacturing facilities were audited by CDSCO and then by US FDA inspectors on behalf of WHO and European Medicines Agency, and this led to the receipt of Emergency Use Listing (EUL) by both WHO and EMA. The entire project from the initiation of facility construction to EUL was achieved in less than 18 months. The overall manufacturing process cycle times were also optimized, and that allowed BioE to achieve the minimum annualized output in excess of 500 million doses of

V. Paradkar (✉) · M. Datla
Biological E Limited, Hyderabad, India
e-mail: Vikram.Paradkar@Biologicale.com

© The Editor(s) (if applicable) and The Author(s), under exclusive license to Springer Nature Singapore Pte Ltd. 2025

R. Swarup et al. (eds.), *At War with the Single Strand*,
https://doi.org/10.1007/978-981-96-8122-8_12

Ad26.COV2.S COVID-19 vaccine. BioE initiated vaccine shipments to Johnson & Johnson global vaccine supply network as well as the COVAX facility in Q1-2022.

12.1 Identification of Strategy for the In-House Development of Vaccine

In the beginning of January 2020, China had announced a novel coronavirus was causing infections leading to a serious respiratory disease, and it was spreading rapidly in the population, and soon patients in many parts of the world started reporting similar infections. This novel coronavirus was similar to the virus, SARS-CoV-1, that had caused the outbreak of Severe Acute Respiratory Syndrome (SARS) in 2002–2004 timeframe. With rapidly mounting cases reported from all corners of the globe, on January 31, 2020, WHO named the disease caused by the SARS-CoV-2 virus as COVID-19 (COronaVIrusDisease 2019) and declared Public Health Emergency of International Concern (PHEIC). As SARS-CoV-2 was a novel virus, global population did not possess any immunity, and the virus was observed to be capable of spreading rapidly through air via the aerosol route. Global scientific and medical professionals quickly realized that an effective vaccine would be the only way to protect the population and get the pandemic under control.

Biological E Limited is a fully integrated vaccine research, development, and manufacturing organization with significant experience in developing a wide variety of vaccines, and we decided to wholeheartedly participate in efforts to provide a safe and effective vaccine against SARS-CoV-2. Lack of large manufacturing facilities that could handle BSL3 microorganisms within the BioE manufacture infrastructure meant we could not manufacture vaccines based on inactivated virus as the antigen, which is the commonly used vaccine platform against other respiratory viruses such as Influenza viruses. Initial evaluation showed significant structural similarities between SARS-CoV-1 and SARS-CoV-2, especially in terms of the presence of spike protein on the virus surface and reported success with the use of the spike protein as the target antigen. Hence, we decided to pursue COVID-19 vaccine based on the adjuvanted protein subunit platform and spike protein as the target antigen. However, the spike protein is highly glycosylated and a very large protein (1273 amino acids) and could be produced either in recombinant mammalian cells or insect cells via baculovirus technology. It was also shown that the SARS-CoV-2 binds to the ACE-2 receptor in human epithelial cells via the Receptor-Binding Domain (RBD), located in the spike protein, and RBD from SARS-CoV-1 was shown to elicit significant and protective immune response in animal studies [1]. RBD, being a much smaller protein (approximately 25 kDa), could be produced in several recombinant microbial systems. As BioE possessed significant expertise and manufacturing facilities for the production of microbial proteins, we decided to focus on RBD protein as the antigen in COVID-19 vaccine development and initiated multiple projects with multiple external collaborators who were developing recombinant microbial strains for RBD protein expression. This was subsequently developed as CORBEVAX (Fig. 12.1).

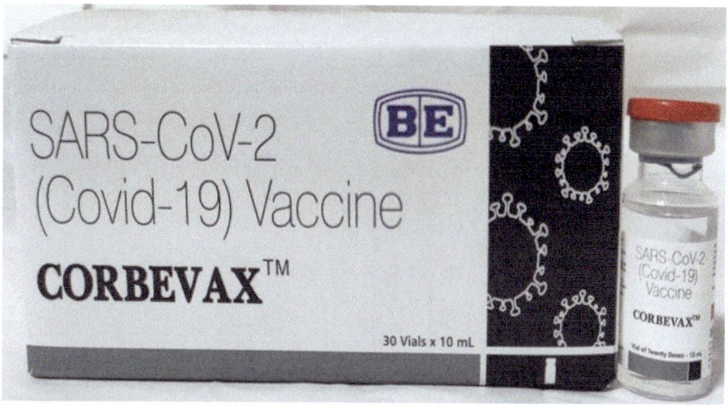

Fig. 12.1 The CORBEVAX vaccine

12.2 Selection of RBD Protein and Candidate Vaccine

Multiple academic labs had started the development of recombinant microbial strains expressing RBD protein, and BioE teams initiated collaboration with these laboratories. Due to severe interruptions in global airline traffic in March–April 2020; the strain shipments to BioE were delayed, and the protein expression, production, and characterization work finally started in early May 2020. We evaluated multiple yeast and E. coli strains for RBD protein expression in shake flasks and pilot-scale fermenters and finalized two RBD protein candidates as suitable for evaluation as antigens. However, RBD being a very small protein, we had to adjuvant it to generate significant immune response based on prior experience. Hence, we contacted multiple adjuvant manufacturers, both commercial suppliers and development organizations, and procured various types of adjuvants such as alum, saponin-based adjuvants, squalene-based adjuvants, and oligodeoxynucleotides (e.g., CpG1018 from Dynavax). We tested the formulation compatibility of RBD protein with the adjuvants and planned multiple mouse studies to evaluate the immunogenicity and safety of various RBD–adjuvant combinations. We developed an in-house ELISA method to determine the concentrations of binding antibodies against the RBD protein but could not execute testing neutralizing antibody titers as these required handling of either live SARS-CoV-2 virus in the BSL3 facility or a pseudovirus-based assay which was available only to academic or research institutions. With BIRAC/DBT help, we could test samples from the mouse immunogenicity study at THSTI for neutralizing the antibody titer against the SARS-CoV-2 strain and with the BMGF support test the nAb titers via Pseudovirus Neutralization Test at Nexelis Lab in Canada. Data from all three laboratories clearly showed significant humoral and cellular immune response when RBD protein was adjuvanted with alum and CpG1018. Interestingly, the magnitude of immune response was much higher when alum and CpG1018 were used together in comparison to the much lower response observed when RBD was adjuvanted with either alum or CpG1018 alone, indicating significant synergistic response of the two co-adjuvants; see Fig. 12.2 for anti-RBD IgG titers and Fig. 12.3 for nAb titers against the SARS-CoV-2 Ancestral strain by MNA. The inclusion of CpG1018, a TLR9 agonist, also provides desired Th1 skew to the humoral and cellular response. Thus, based on the

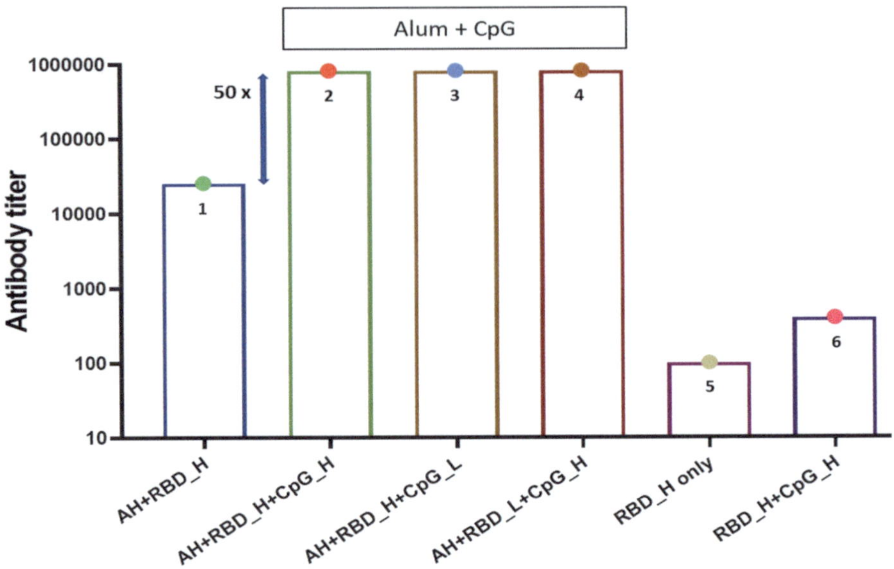

Fig. 12.2 Anti-RBD IgG titer in mouse sera post two-dose administration of various RBD containing vaccine formulations
AH: Aluminum hydroxide
RBD: Receptor-Binding Domain protein as antigen. "H" and "L" designate for higher and lower concentrations of RBD in the formulation
CpG: CpG1018 as adjuvant. "H" and "L" designate for higher and lower concentrations of RBD in the formulation

significant and robust immunogenicity observed in the animal studies, we selected RBD protein with alum and CpG1018 adjuvants as candidate vaccine for clinical development. Based on the productivity observed in pilot-scale runs and excellent compatibility with the formulation studies, we selected the RBD protein expressed in the recombinant Pichia pastoris strain that was developed by researchers at Baylor College of Medicine, Texas Children's Hospital.

In order to initiate clinical development, we had to conduct crucial tasks related to vaccine manufacturing under GMP conditions along with detailed product characterization (RBD protein and vaccine formulation) as well as demonstrate appropriate product stability. Immediately after demonstration of excellent immunogenicity and safety in the mouse model, we decided to scale-up the manufacturing process of RBD protein in GMP suites in commercial-scale fermenters and completed the manufacturing of RBD antigen for clinical studies. A vaccine formulation consisting of maximum concentration of RBD protein, alum, and CpG1018 was finalized, and an acute and repeat-dose safety study was initiated in rats at the BioE-Animal House. In parallel, a more comprehensive and extensive repeat-dose safety study in rats and rabbits was initiated at an external animal testing facility. As per established guidelines, we submitted Phase I/II clinical trial application in October 2020 which included all the CMC information related to RBD protein production consistency,

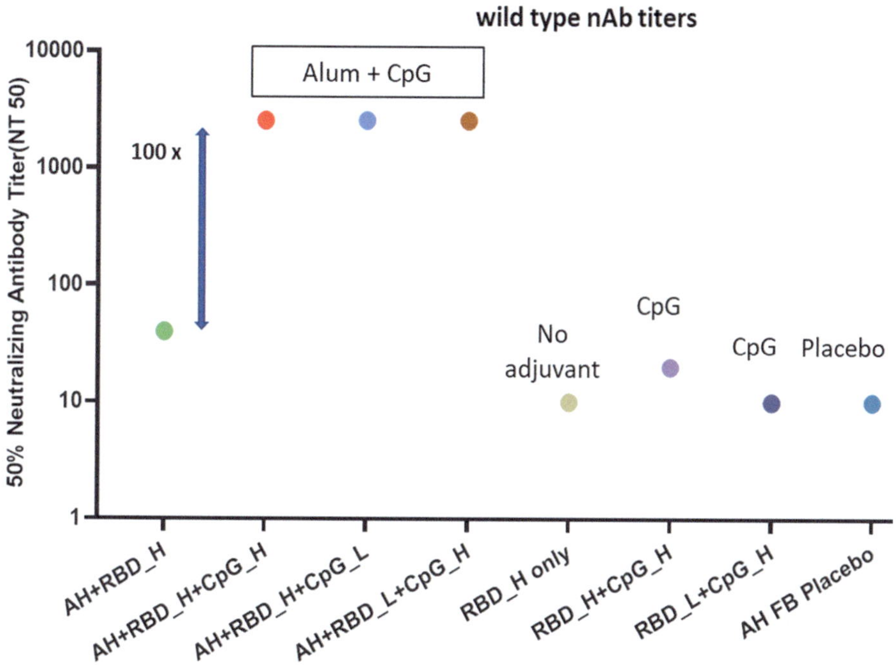

Fig. 12.3 nAb titers against SARS-CoV-2 Ancestral strain in mouse sera samples post two-dose administration of various RBD containing vaccine formulations
Pooled sera from each vaccine formulation cohort was tested for nAb titers
AH: Aluminum hydroxide
RBD: Receptor-Binding Domain protein as antigen. "H" and "L" designate for higher and lower concentrations of RBD in the formulation
CpG: CpG1018 as adjuvant. "H" and "L" designate for higher and lower concentrations of RBD in the formulation

detailed characterization on purity, structure and biological activity, stability, as well as the production consistency, characterization, and stability of vaccine formulation. This was supplemented with the excellent safety data from the animal studies and the immunogenicity data showing consistent immune response post two-dose vaccination regimen.

12.3 Phase I/II Clinical Trial for Selection of Vaccine Formulation

A detailed review of BioE Phase I/II clinical trial protocol and vaccine CMC-related information was conducted in the Subject Expert Committee and CDSCO teams, and approval was granted to initiate the trials at six sites across India. BIRAC provided significant funding that covered the preclinical vaccine evaluation and conduct of Phase I/II clinical trials as part of National Biopharma Mission. The Phase

I/II study included 360 adult RT-PCR negative and SARS-CoV-2 seronegative subjects who were administered two doses of vaccine 28 days apart via intramuscular route. A total of four different vaccine formulations were evaluated for safety and immunogenicity, as summarized in Table 12.1. Sera and whole-blood samples for PBMC stimulation were collected pre-vaccination, Day 28 post first-dose, Day 42, and Day 56, i.e., 14 and 28 days after the second dose. The sera samples were tested for anti-RBD IgG concentration, anti-RBD IgG1 and IgG4 subclass titers (to assess Th1 vs Th2 humoral response), and nAb titers against the Ancestral SARS-CoV-2 strain by the MNA method at THSTI and against pseudovirus at Nexelis Lab, Canada. The nAb titer testing was executed with support from BIRAC and BMGF. The PBMCs from whole-blood samples were stimulated with SARS-CoV-2 peptides, and the secreted cytokines were tested by ELISA to determine the relative Th1 versus Th2 skew of the cellular immune response. All four formulations generated good immune response in terms of significant increase in the anti-RBD IgG titers post two doses, and good nAb titers were observed in both PSVNT (see Fig. 12.4) and MNA assays (see Fig. 12.5) [2]. The highest immune responses observed in the "B" formulation were comparable to the GMT observed for the Human Convalescent Sera panel GMTs. The high ratio of IgG1 titers versus IgG4 titers and the significant increase in interferon-gamma in comparison to interleukin-4 cytokine secretion by stimulated PBMCs indicates that both humoral and cellular immune responses showed desired Th1 skew. During the Phase I/II clinical trial execution, the Beta strain of SARS-CoV-2 was observed to possess the highest immune-escape properties, and hence nAb titers were tested against the pseudovirus expressing the Beta strain spike protein at Nexelis Lab. This testing showed significant nAb titers against the Beta strain with a minimal drop from Ancestral strain titers (see Fig. 12.6). Most of the other COVID-19 vaccines had shown significant reduction in nAb titers against the Beta strain.

The Phase I/II trials also demonstrated the crucial role CpG1018 adjuvant plays in the generation of robust immune response, with a significantly higher immune response observed in formulations containing 500 mcg of CpG vs 250 mcg of CpG per dose for the same RBD content of 25 mcg per dose. In order

Table 12.1 Composition of vaccines used in Phase I/II study

S. No.	Process function	RPN value	Risk level
1	Facility design and construction	24	Minor
2	Biosafety	20	Minor
3	HVAC system nonbiocontainment area	12	Minor
4	HVAC system biocontainment area	24	Minor
5	Utilities (purified water/WFI/pure steam/compressed air/CO_2 gas)	20	Minor
6	Electrical	20	Minor
7	Network communication	20	Minor
8	Environmental monitoring program	20	Minor
9	Personnel	20	Minor
10	Equipment	16	Minor

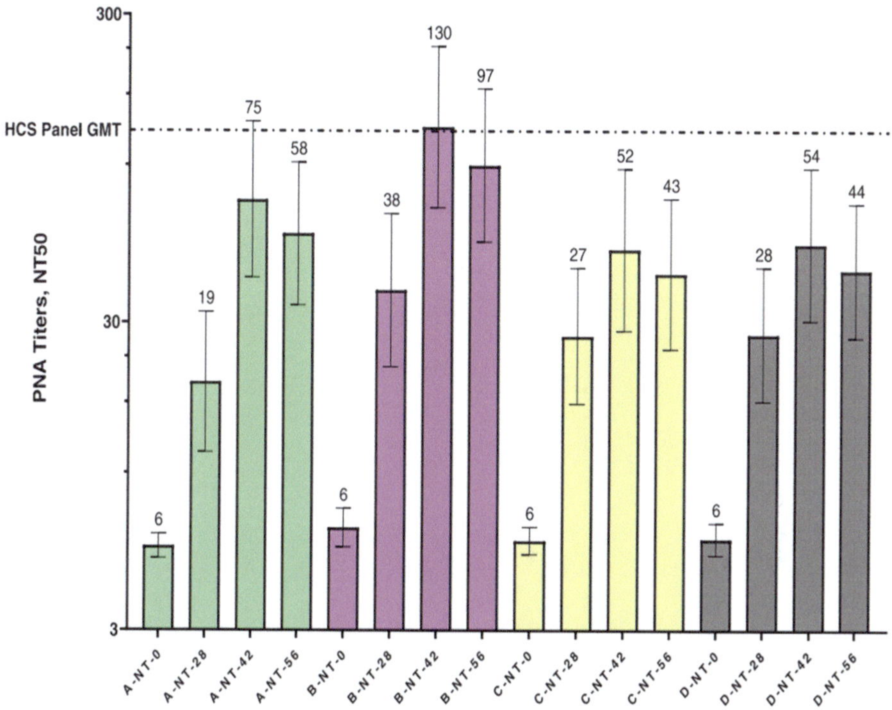

Fig. 12.4 Pseudovirus neutralization titer GMTs (and 95% CI) at Days 0, 28, 42, and 56 time-points from Phase I/II study in four vaccine formulations A, B, C, D containing different quantities of RBD and CpG1018
HCS Panel: Pool of Human Convalescent Sera samples collected from 270 subjects with varying severity of disease in Canada in early 2020

to enhance the immunogenicity potential of the vaccine, we decided to further increase the CpG1018 content in the formulation to the highest capacity of the formulation with almost quantitative adsorption. This formulation development work resulted in increasing the CpG1018 content to 750 mcg per dose.

The Phase I/II clinical trial data and the formulation development data on increasing the CpG1018 content was subsequently shared with SEC and CDSCO. In order to demonstrate the safety and benefit of enhanced immunogenicity of the higher CpG1018-containing formulation (termed as formulation "E"), the committees recommended to conduct a Phase II trial. This trial was conducted in 100 adult subjects with the same two-dose regimen. This trial showed that the formulation had the same excellent safety profile and demonstrated significant increase in both binding-IgG and nAb titers in both PSVNT and MNA assays (see Figs. 12.7, 12.8, and 12.9) which met our goals for an optimum COVID-19 vaccine [2]. The optimum formulation also demonstrated significant nAb titers against both Delta and Beta strains of SARS-CoV-2 in the MNA testing conducted at THSTI using virus strains isolated in India. The Phase I/II trials leading to the finalization of the candidate COVID-19 vaccine were conducted between Nov 2020 and July 2021 at multiple sites in India.

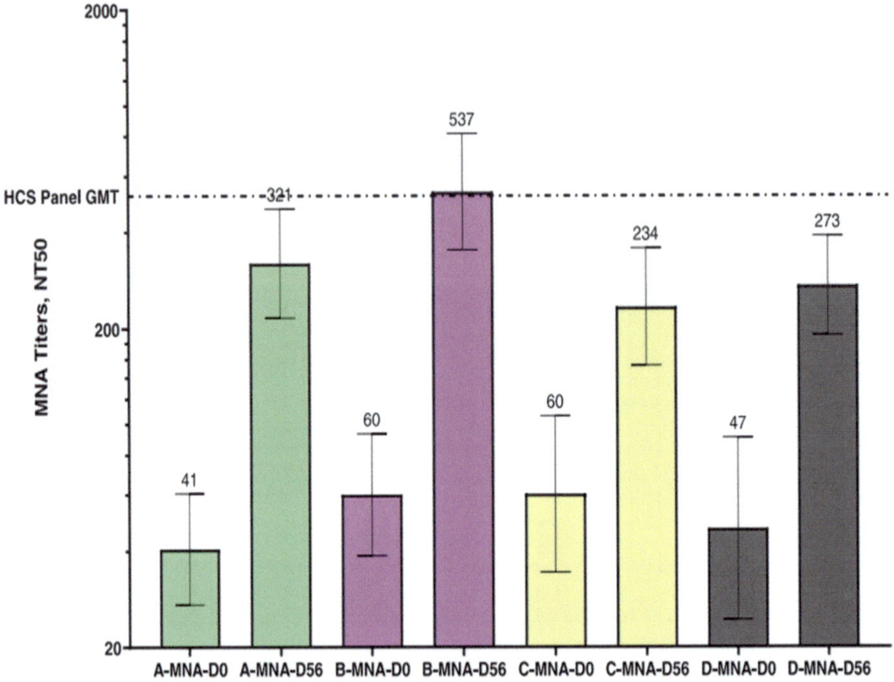

Fig. 12.5 MicroNeutralization titer GMTs (and 95% CI) against the Ancestral strain at Day 0 and Day 56 time-points from Phase I/II study in four vaccine formulations A, B, C, D containing different quantities of RBD and CpG1018
HCS Panel: Pool of Human Convalescent Sera samples collected from 34 subjects with severe COVID19 disease in UK in early 2020

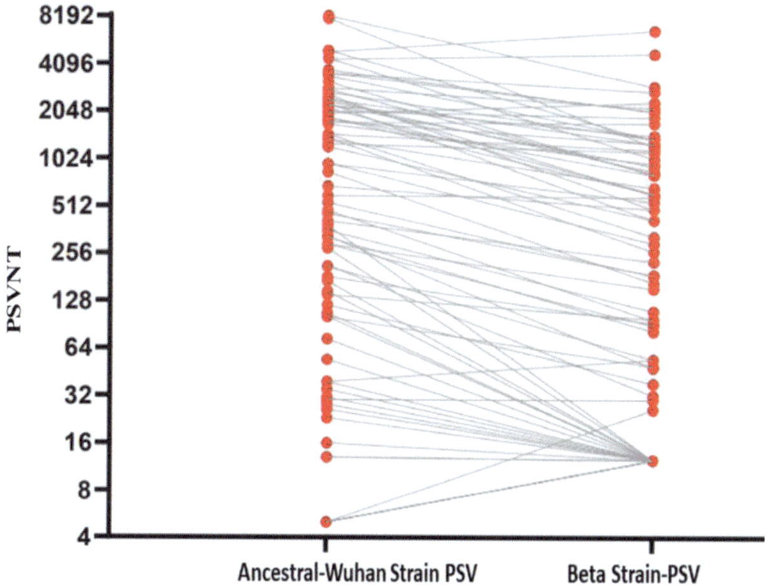

Fig. 12.6 Pseudovirus neutralization titers against Ancestral-Wuhan strain and Beta strain of SARS-CoV-2 reported in a paired format
Sera samples collected post two-dose vaccination of various formulations used in Phase I/II study. Average fold-reduction in the pseudovirus nAb titer from Ancestral to Beta strain was 2.25-fold

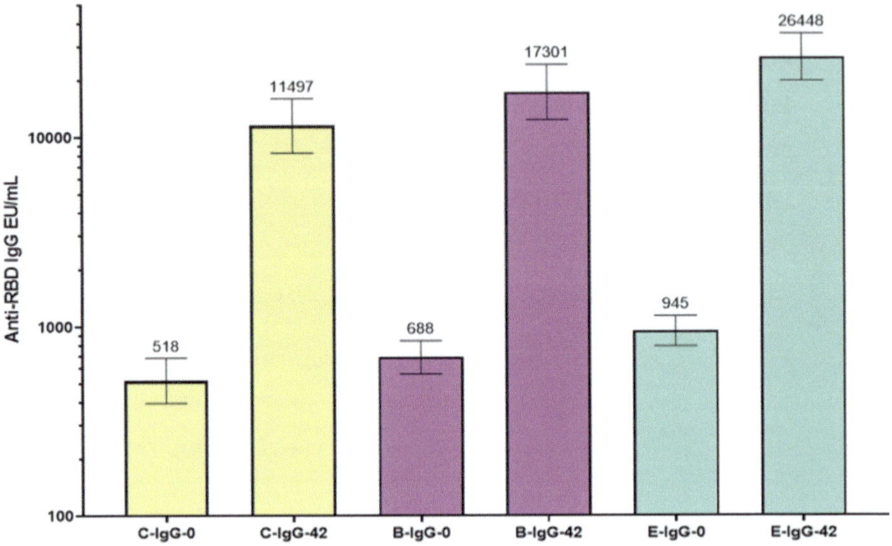

Fig. 12.7 Anti-RBD IgG concentration GMC (95% CI) comparison between formulations C, B, and E from Phase I and II clinical trials
C formulation: 250 microgram CpG1018 per dose
B formulation: 500 microgram CpG1018 per dose
E formulation: 750 microgram CpG1018 per dose
All formulations contained 25 microgram RBD per dose
D-0: Day 0; pre-vaccination
D-42: Day 42; post two-dose vaccination

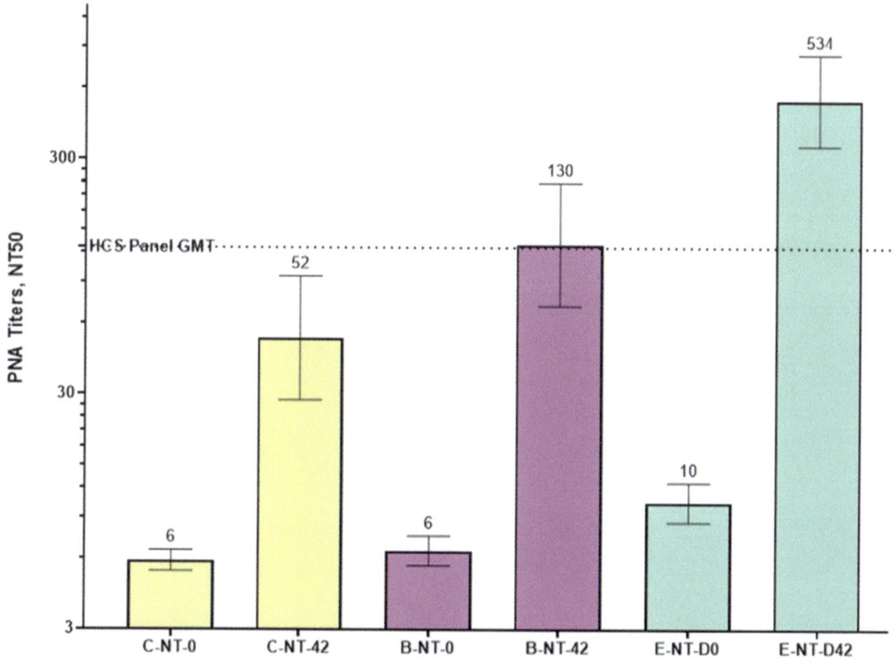

Fig. 12.8 Pseudovirus neutralization titer GMTs (95% CI) comparison concentration GMC (95% CI) between formulations C, B, and E from Phase I and II clinical trials
C formulation: 250 microgram CpG1018 per dose
B formulation: 500 microgram CpG1018 per dose
E formulation: 750 microgram CpG1018 per dose
All formulations contained 25 microgram RBD per dose
D-0: Day 0; pre-vaccination
D-42: Day 42; post two-dose vaccination

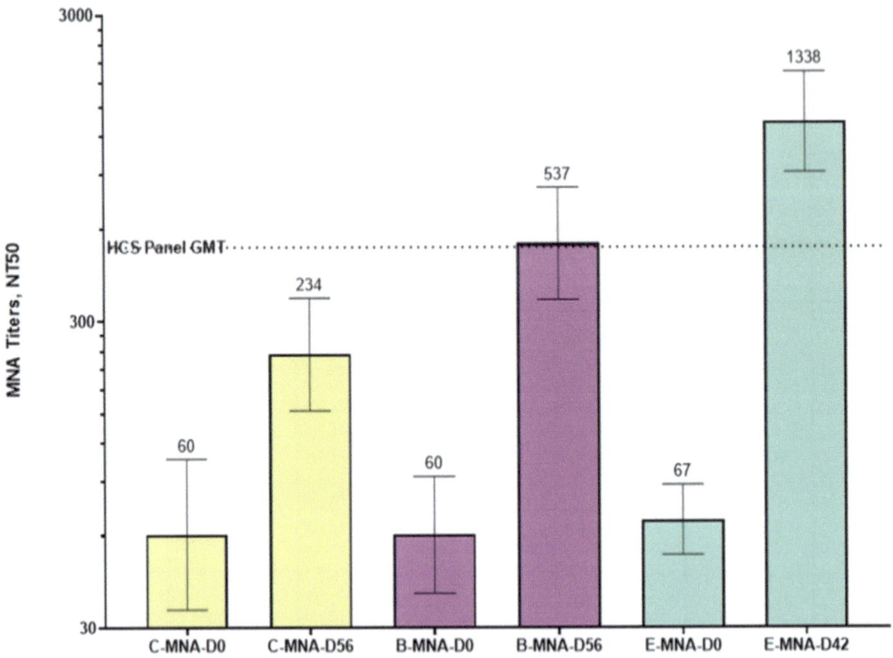

Fig. 12.9 MicroNeutralization titer GMTs (and 95% CI) against the Ancestral strain comparison between formulations C, B, and E from Phase I and II clinical trials
C formulation: 250 microgram CpG1018 per dose
B formulation: 500 microgram CpG1018 per dose
E formulation: 750 microgram CpG1018 per dose
All formulations contained 25 microgram RBD per dose
D-0: Day 0; pre-vaccination
D-42: Day 42; post two-dose vaccination

12.4 CORBEVAX™ Pivotal Clinical Trials for Adults and First EUA

As the clinical development of the RBD vaccine was advancing, we decided to scale up the RBD protein manufacturing process to eventual commercial-scale in the large Drug Substance suite that was constructed for microbial protein production. CEPI supported the scale-up projects for both the RBD-Drug Substance and vaccine formulation processes. CEPI also worked with the CpG1018 manufacturer, Dynavax, to increase the production capacity of the adjuvant as it was under evaluation in multiple COVID-19 vaccines. The CEPI support for manufacturing scale-up and adjuvant availability helped BioE's efforts to qualify and validate the vaccine production capacity of approximately 100 million doses per month in view of expectation of supplying a large quantity of vaccine in a very short period of time. Both the RBD protein and vaccine formulation processes were successfully scaled up and validated in Q2'21. The products were fully characterized, demonstrated to be comparable to the products used in the early phase clinical trials, and also showed appropriate stability profiles at the recommended storage temperature. The product scale-up CMC data on process and product-quality consistency, extended characterization, and stability coupled with data from Phase I/II clinical trials was submitted to CDSCO and SEC as part of BioE's application to conduct pivotal trials for vaccine approval. After multiple reviews, the committees recommended conducting two pivotal clinical trials in adult population. COVID-19 vaccines were initially approved based on conducting classical vaccine efficacy studies in seronegative subjects; in these efficacy studies, the vaccinated cohorts showed significant reduction in COVID-19 infection (serious infection and hospitalization) in comparison with the placebo cohorts. However, by August 2021, vaccination campaigns were already underway in India, and the percentage of seronegative adults was observed to be very low (10–20%) in the most recent clinical studies. Thus, it was not possible to conduct a classical placebo-controlled efficacy study for product approval. With the recommendations from the SEC, BioE finalized a pivotal clinical trial with a primary end-point related to the demonstration of superior immunogenicity of CORBEVAX™ in comparison to the approved vaccine COVISHIELD™. In addition to this trial, we also planned to conduct another Phase III trial to specially evaluate safety and immunogenicity in elderly population. These clinical trials also ensured minimum 3000 adult subjects received two doses of CORBEVAX™ to generate a sufficiently large product safety database. In November 2020, BIRAC launched the ambitious COVID SURAKSHA mission to accelerate COVID-19 vaccine development in India, and BioE received significant funding to conduct multiple pivotal clinical trials of CORBEVAX™ at more than 25 clinical sites in India. DBT and CEPI also provided significant support by prioritizing the crucial nAb titer testing at THSTI for all the pivotal clinical trials.

The adult clinical trials were conducted in the August–October 2021 timeframe, and CORBEVAX™ met all the primary and secondary end-points to demonstrate superior immunogenicity in comparison to COVISHIELD™ [3]. The superiority was demonstrated in terms of significantly higher GMTs for nAb titers against both Ancestral strain and the Delta strain of SARS-CoV-2 in CORBEVAX™ recipients in comparison to COVISHIELD™ recipients, as summarized in Table 12.2. The magnitude of the cellular immune response, as indicated by the number of PBMCs secreting cytokine interferon-gamma post stimulation with spike protein peptides, was also higher in CORBEVAX™ recipients in comparison to COVISHIELD™ recipients, as shown in Fig. 12.10. Both CORBEVAX™ and COVISHIELD™ vaccination cohorts showed good safety profile, and the CORBEVAX™ cohort had 50% lower adverse events in comparison to COVISHIELD™. The clinical trial conducted in elderly population also demonstrated excellent safety and immunogenicity in terms of both humoral and cellular immune response and nAb titers against Ancestral, Beta, and Delta strains.

While the two adult Phase III studies were underway, we also completed two crucial animal studies to evaluate CORBEVAX™ safety and efficacy. UKHSA had established an excellent non-human primate model for SARS-CoV-2 virus infection, and with CEPI support, they conducted a NHP challenge study in which the

Table 12.2 Comparison of neutralization titers (95% CI) post CORBEVAX vaccination and COVISHIELD vaccination in Phase III study

SARS-COV-2 strain used in MicroNeutralization Assay	Vaccine	Day-0 GMT (95% CI)	Day-42 GMT (95% CI)
Ancestral strain	CORBEVAX	85	2123
		(75–96)	(1801–2514)
	COVISHIELD	75	1833
		(65–86)	(1632–2089)
Delta strain	CORBEVAX	ND	874
			(724–1055)
	COVISHIELD	ND	562
			(482–657)

Titers tested against Ancestral strain and Delta strain of SARS-CoV-2 by MicroNeutralization Assay

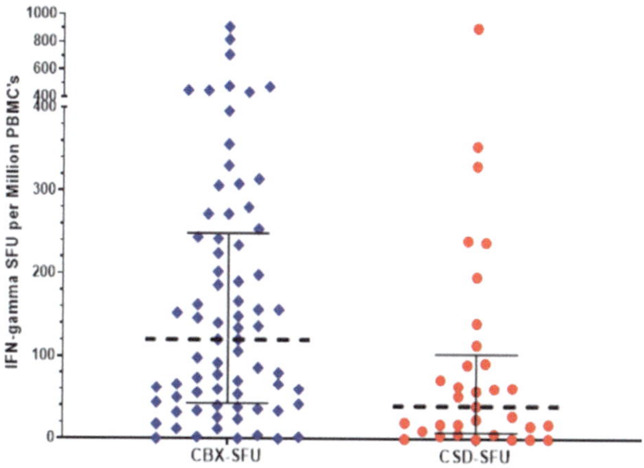

Fig. 12.10 Comparison of the magnitude of cellular immune response in CORBEVAX™ recipients vs. COVISHIELD™ recipients from Phase III clinical trial
Interferon-gamma secreting PBMCs observed as SFU (Spot Forming Unit) in ELISPOT assay post-stimulation with SARS-CoV-2 spike protein peptides. PBMCs collected on Day 42 post two-dose vaccine administration
CBX: CORBEVAX recipients; median PBMC count 120 (IQR: 43–249)
CSD: COVISHIELD recipients; median PBMC count 40 (IQR: 8–103)

NHPs were first immunized with two or three doses of CORBEVAX™ and then subsequently challenged with SARS-CoV-2 virus. Post challenge, the degree of infection was assessed in terms of viral RNA load in nasal and throat passage and lung extracts. The severity of the infection was assessed via lung CT scan and examination of tissues such as heart. CORBEVAX™-vaccinated NHPs showed almost a complete absence of viral RNA in various tissue samples, and minimal disease severity was observed in the lung CT scan and tissue examination. In comparison, NHPs that received placebo vaccine had very high viral RNA load and higher disease severity in lung tissue. This study showed the significant potential of CORBEVAX™ to prevent SARS-CoV-2 infection and reduce disease severity in NHPs. We also conducted a specific study to assess the Developmental And Reproductive Toxicity (DART) in female rats. A total of five doses of CORBEVAX™ were administered to female rats over approximately 50 days duration (two doses prior to conception, two doses during gestation, and one dose post birth during lactation). Both the placebo and CORBEVAX™ cohorts showed no impact on any reproductive parameters, and mothers and pups were healthy during the course of the study. Significant immunogenicity was reported in maternal sera samples, and most importantly, significant transmission of maternal anti-RBD IgG transfer was observed to the fetus and pups. This study serves as an important justification for the potential use of CORBEVAX™ in pregnant females and females of child-bearing age.

The data from the adult clinical trials and additional animal studies were submitted to CDSCO and discussed extensively in SEC meetings in December 2021, and as a result, CDSCO provided Emergency Use Authorization (EUA) for CORBEVAX™ for adult population on December 28, 2021.

12.5 CORBEVAX™ Pivotal Clinical Trials in Pediatric Subjects

The excellent safety profile of CORBEVAX™ in adult clinical trials and lack of sufficient vaccination options for pediatric population prompted us to accelerate the clinical evaluation of CORBEVAX™ in the age group of 5 to <18. A Phase II/III clinical trial application was submitted for conducting this study, Phase II in a small group of subjects in 12 to <18 years age group, and Phase III in the two cohorts viz., 5 to <12 and 12 to <18 years. The study was initiated in November 2021 and completed in February 2022. This study was partly funded by BIRAC grant as part of the COVID SURAKSHA mission. CORBEVAX™ vaccine formulation and dosing regimen for the pediatric population was kept the same as that employed for adult studies. This study also demonstrated the excellent safety profile of CORBEVAX™ in both the pediatric age cohorts. Immunogenicity assessment showed comparable immune response in the pediatric population to the adult cohort in terms of binding-IgG titers and nAb titers measured against the Ancestral SARS-CoV-2 strain by the MNA method [4]. The immunogenicity data is summarized in Table 12.3. The high ratio of anti-RBD IgG1 GMT to anti-RBD IgG4 GMT indicated desired Th1 skew to the humoral immune response, as observed in the adult population. Cellular immune response was evaluated by stimulating PBMCs in either an isolated format or in whole-blood samples with SARS-CoV-2 peptides. The inferferon-gamma-secreting PBMCs were quantified by the ELISPOT method, and cytokines secreted from whole-blood stimulation were tested by ELISA. The cellular immune response in the pediatric population was also similar to the adult experience and also demonstrated the desired Th1 skew. Based on this clinical trial data, CDSCO provided Emergency Use Authorization for CORBEVAX™ for use in the 12 to <18 years age group on February 21, 2022 and for use in the 5 to <12 years age group on April 26, 2022. Based on this approval, Government of India initiated pan-India campaign of vaccination of children of ages 12, 13, and 14 from March 16, 2022.

Table 12.3 Summary of immunogenicity parameters from the pediatric Phase II/III study and comparison with prior adult study

	Age cohort	Day-0	Day-42
Anti-RBD IgG concentration, GMC EU/mL	Children 12 to <18	939	18,049
	Children 5 to <12	964	26,802
	Adults: 18–55	945	26,448
Neutralization titer against Ancestral SARS-CoV-2 strain, GMT	Children 12 to <18	50	1099
	Children 5 to <12	44	1148
	Adults: 18–55	67	1338

Table 12.4 Comparison of immunogenicity parameters in CORBEVAX heterologous booster dose recipients in comparison with placebo recipients in Phase III clinical trial

		Primary vaccination COVISHIELD™		Primary vaccination COVAXIN™	
		Day-0	Day-28	Day-0	Day-28
Anti-RBD IgG GMC; EU/mL	CORBEVAX™	19,825	34,975	20,182	51,248
	Placebo	23,235	19,564	24,311	30,392
Neutralization titers against SARS-CoV-2 Ancestral strain	CORBEVAX™	1143	6317	834	5622
	Placebo	1758	1877	1193	2366
Pseudovirus neutralization titers against Omicron B.1 strain	CORBEVAX™	14.9	78.8	11.2	45.2
	% Responders	43%	91%	33%	75%

12.6 CORBEVAX™ Clinical Trials for Use as a Heterologous Booster Dose Vaccine

Most of the COVID-19 vaccines used globally showed significant waning of immune response after approximately 6 months of completion of primary vaccination. Similar data was also reported in India in adults who had received either COVISHIELD™ or COVAXIN™ in two-dose primary vaccination in 2021 [5]. Hence, we evaluated CORBEVAX™ as a heterologous booster dose vaccine in a Phase III study in adults who had previously received the two-dose regimen of either COVISHIELD™ or COVAXIN™ in comparison with a placebo booster dose [6]. This trial showed the single heterologous booster dose administration of CORBEVAX™ was extremely safe as indicated by the percentage of subjects reporting adverse events were similar between the CORBEVAX™ arm and placebo arm, and most of the adverse events were mild, and no atypical or grade-3 adverse events were observed. The CORBEVAX™ recipients showed significant increase in nAb titers and binding-IgG titers in comparison to the placebo recipients as summarized in Table 12.4. This data shows the clear benefit of using CORBEVAX™ as a heterologous booster dose in adults to enhance the immune response that had waned post 6–9 months from the primary vaccination with either COVISHIELD™ or COVAXIN™. Based on this Phase III clinical trial data, CDSCO provided Emergency Use Authorization on June 4, 2022 for use of CORBEVAX™ as a heterologous booster to adults who had received the two-dose primary vaccination of COVISHIELD™ or COVAXIN™.

12.7 CORBEVAX Supplies and Pan-India Campaign

During the clinical development phase of CORBEVAX™, based on the promising results, Government of India—Ministry of Health placed advance purchase orders, and BioE initiated manufacturing of CORBEVAX™ toward the stock-pile of the product in between August 2021 till January 2022. In March and April 2022, BioE supplied approximately 100 million doses of CORBEVAX™ for use in the pediatric vaccination campaign. CORBEVAX™ was supplied to Government of India at less than Rs 150 per dose, which represents the most affordable COVID-19 vaccine globally, fulfilling the promise made by BioE to BIRAC, CEPI, and BMGF regarding the supply of large quantities of safe, effective, and affordable vaccine. Children of 12, 13, and 14 years of age received CORBEVAX™ exclusively, and approximately 75 million doses of CORBEVAX™ were administered to children, with approximately 33 million children receiving both doses of the vaccine. Government of India also included CORBEVAX™ as a heterologous booster dose vaccine in adults from August 2022, and approximately ten million adults have received CORBEVAX™ as a booster dose. Excellent safety profile was observed in both pediatric and adult Pan-India vaccination campaigns as was expected based on the clinical trial experience. This is indicated by the extremely low number of vaccine-related adverse events reported to the AEFI committee from CORBEVAX™ recipients.

12.8 Variant Vaccine Development

COVID-19's pandemic phase is now over, but the SARS-CoV-2 virus keeps on mutating and thus still able to circulate globally. The mutations in the spike protein have resulted in significant immune-escape potential which results in people getting infected even after vaccination or prior infection. On May 16, 2023, WHO TAG-CO-VAC group published guidelines recommending the development of COVID-19 vaccines targeting the SARS-CoV-2 variant XBB.1.5 for use in the near future. This variant-specific vaccine is expected to be important for elderly and immune-compromised individuals to reduce the risk of severe disease. BioE has started the development of RBD antigen that matches the mutations present in XBB.1.5, and the candidate vaccine is expected to be available in the near future for clinical evaluation and the subsequent wider distribution.

12.9 Conclusions

Biological E developed completely in-house and fully indigenous novel CORBEVAX™ vaccine in a short duration and supplied large quantities of extremely safe, effective, and highly affordable vaccine for Pan-India deployment. This development was supported by significant grants and technical help from BIRAC and CEPI and BMGF organizations. This vaccine development has also provided valuable experience which will be very important to address future pandemics or outbreaks.

Competing Interest The authors have no conflicts of interest to declare that are relevant to the content of this chapter.

References

1. Chen W-H, Lanying D, Chag SM, Ma C, Tricoche N, Tao X, Seid CA, Hudspeth EM, Lustigman S, Tseng C-TK, Bottazzi ME, Hotez PJ, Zhan B, Jiang S. Yeast-expressed recombinant protein of the receptor-binding domain in SARS-CoV spike protein with deglycosylated forms as a SARS vaccine candidate. Hum Vaccin Immunother. 2014;10(3):648–58.
2. Thuluva S, Paradkar V, Gunneri SR, Yerroju V, Mogulla R, Turaga K, Kyasani M, Manoharan SK, Medigeshi G, Singh J, Shaman H, Singh C, Rao V. A evaluation of safety and immunogenicity of receptorbinding domain-based COVID-19 vaccine (Corbevax) to select the optimum formulation in open-label, multicentre, and randomised phase-1/2 and phase-2 clinical trials. Lancet eBioMedicine. 2022;2022(83):104217.
3. Thuluva S, Paradkar V, Gunneri SR, Yerroju V, Mogulla R, Suneetha PV, Turaga K, Kyasani M, Manoharan SK, Adabala S, Javvadi AS, Medigeshi G, Singh J, Shaman H, Binayke A, Zaheer A, Awasthi A, Singh C, Venkateshwar Rao A, Basu I, Kumar KAA, Pandey AK. Immunogenicity and safety of Biological E's CORBEVAX™ vaccine compared to COVISHIELD™ (ChAdOx1 nCoV-19) vaccine studied in a phase-3, single blind, multicentre, randomized clinical trial. Hum Vaccin Immunother. 2023;19(1):2203632.
4. Thuluva S, Paradkar V, Gunneri SR, Yerroju V, Mogulla RR, Suneetha PV, Turaga K, Kyasani M, Manoharan SK, Adabala S, Javvadi AS, Medigeshi G, Singh J, Shaman H, Binayke A, Zaheer A, Awasthi A, Narang M, Nanjappa P, Mahantshetti N, Garg BS, Pandey AK. Safety, tolerability and immunogenicity of Biological E's CORBEVAXTM vaccine in children and adolescents: a prospective, randomised, doubleblind, placebo controlled, phase-2/3 study. Vaccine. 2022;40(2022):7130–40.
5. Selvavinayagam ST, Yong YK, Tan HY, Zhang Z-Y, Subramanian G, Rajeshkumar M, Vasudevan K, Jayapal P, Narayanasamy K, Ramesh D, Palani S, Larsson M, Shankar EM, Raju S. Factors associated with the decay of anti-SARS-CoV-2 neutralizing antibodies among recipients of an adenoviral vector-based AZD1222 and a whole-virion inactivated (BBV152) vaccine in Chennai, India: a prospective, longitudinal, cohort study. MedRxiv. 2022, February 2022.
6. Thuluva S, Paradkar V, Gunneri SR, Yerroju V, Mogulla R, Suneetha PV, Turaga K, Binayke A, Zaheer A, Awasthi A, Virkar R, Narang M, Nanjappa P, Mahantshetti N, Garg BS, Ravi MR. Immunogenicity and safety of Biological E's CORBEVAX™ vaccine as a heterologous booster dose in adult volunteers previously vaccinated with two doses of either COVISHIELD™ or COVAXIN: a prospective double-blind randomised phase III clinical study. MedRxiv, January 2023.

Make in India Vaccines: A True Story of Atmanirbhar Bharat—Hester Biosciences Limited's Contribution Toward Combating COVID-19

13

Priya Gandhi and Kajal Patel

Hester Biosciences Limited is one of India's leading poultry and animal vaccine manufacturing company, since 1997. Hester operates in two broad product segments:

1. Vaccines
2. Health products
   1. It is the world's largest manufacturer and supplier of PPR vaccine, having approximately 75% of the world market.
   2. It is now the second largest poultry vaccine manufacturer, with approximately 35% market share in India.

Hester's vaccine capabilities include multiple platforms such as Chick Embryo Origin, Continuous Cell line, Tissue Culture, and Fermentation-based live as well as inactivated vaccines.

Hester recently launched the Petcare Division with a view to focus on the needs of companion animals.

Hester recognizes the vision of ONE HEALTH and works on improving the health of animals by enabling better health for human beings (https://www.hester.in/).

13.1 Background on the Project

In 2020, the world plunged into an unprecedented global crisis due to the COVID-19 pandemic. Many institutes, worldwide, embarked on a mission to develop the vaccine against COVID-19 to immunize and protect mankind.

P. Gandhi (✉) · K. Patel
Hester Biosciences Limited, Ahmedabad, Gujarat, India
e-mail: priya.gandhi@hester.in

© The Editor(s) (if applicable) and The Author(s), under exclusive license to Springer Nature Singapore Pte Ltd. 2025
R. Swarup et al. (eds.), *At War with the Single Strand*,
https://doi.org/10.1007/978-981-96-8122-8_13

Two Indian companies were granted Emergency Use Authorization (EUA) by the Government of India to manufacture the COVID-19 vaccine to help the government to achieve its objective to vaccinate and immunize the entire eligible population in India and be a supplier of the said vaccine to the world.

To meet the target and to augment the capacity of existing manufacturers, the government invited the public and the private sectors to come forward and help increase the manufacture and supply of vaccines.

To achieve this target, Gujarat Biotechnology Research Centre (GBRC) & Hester formed a consortium in the name of Gujarat COVID-19 Vaccine Consortium (GCVC) to manufacture the drug substance for COVID-19 vaccine. GCVC then entered into a licensing agreement with Bharat Biotech International Limited (BBIL) to manufacture the drug substance for Covaxin at Hester through a greenfield project.

Hester constructed a multipurpose BSL-III laboratory to manufacture the Covaxin drug substance.

13.2 About the Facility

The facility has three buildings—manufacturing, utility, and warehouse.

The manufacturing building comprises four levels:

Minus 1: Primary heat treatment and decontamination of effluent from the biocontainment area

Ground: Manufacturing area (cell culture activities, biocontainment processes, purification processes, and IPQC testing)

Ground + 1: Technical area housing HVAC systems (containment and others), clean utility generation system, utility piping, and electrical cables and tray

Ground + 2: General QC, HVAC system and utility services for QC, electrical panels, and administrative office

Facility qualification has been performed as per guidelines.

Guidelines for the establishment of containment facilities: Biosafety levels 2 (BSL-2) and 3 (BSL-3) and certification of BSL-3 facility 2020.

Laboratory biosafety manual 3rd edition (2004), World Health Organization, Geneva, 2004.

13.3 Major Features

1. Biocontainment areas are in isolated zones within a building, as per the "Box in Box" principle.
2. Two-stage Bag-In- Bag-Out (BIBO) HEPA filters are placed.
3. Zero leakage dampers are placed.
4. Pressure hold test is performed one by one as a group of rooms which are related to one common AHU and finally the entire biocontainment area. Pressure drops less than 125 Pa (half of original 250 Pa) in 20 min.
5. Vaporized hydrogen peroxide pass box is provided.
6. The periphery of biocontainment is controlled by airtight doors.
7. All the process area drains and shower drains in the biocontainment areas are connected to the Kill tank through the SS pipeline with NRV.
8. All electrical power cables and network cables which relate to the biocontainment process areas' entry points are covered with Roxtec glands.

13.4 Risk Assessment

Risk assessment is performed to identify the potential risks involved in "manufacturing of viral vaccine drug substance in (BSL III) facility." The RPN score at each step is classified as Minor. Adequate controls are in place to mitigate the identified risks. Hence, it is concluded that for "manufacturing of viral vaccine drug substance (Whole Virion, Inactivated Coronavirus (SARS-CoV-2) Vaccine bulk) in (BSL III) facility" does not pose any potential risk (mentioned in Table 13.1).

Table 13.1 Risk level of BSL III facility

S. No.	Process function	RPN value	Risk level
1	Facility design and construction	24	Minor
2	Biosafety	20	Minor
3	HVAC system non-biocontainment area	12	Minor
4	HVAC system biocontainment area	24	Minor
5	Utilities (purified water/WFI/pure steam/compressed air/CO_2 gas)	20	Minor
6	Electrical	20	Minor
7	Network communication	20	Minor
8	Environmental monitoring program	20	Minor
9	Personnel	20	Minor
10	Equipment	16	Minor

13.5 Current Scenario

Due to the widespread vaccination by the Government of India against the COVID disease, the COVID-19 epidemic has been fully brought under control. New COVID-19 cases are few, the demand for COVID-19 vaccines has substantially come down, and the vaccine manufacturers are sitting on a pile of unsold vaccines.

In such a situation, Hester is looking at an alternative use of the facility.

13.6 Alternate Use(s) of the Facility

A. Keeping the facility in an operating condition is important to achieve the strategic intent behind creating the capacity.
B. This BSL-3 drug substance manufacturing facility is designed to handle the COVID-19 vaccine. However, the same can be re-purposed for manufacturing other human vaccines. Based on a preliminary study, Hester has evaluated the following options for the alternate use of the facility:
   (a) Drug substance for the following human vaccines can be manufactured with minor changes in the manufacturing facility and with the provided Vero cell substrate:
      (i) Rabies vaccine
      (ii) Japanese Encephalitis vaccine
      (iii) Inactivated polio vaccine
      (iv) Dengue vaccine
   (b) Drug substance for the following human vaccines can be manufactured with minor changes in the manufacturing facility but with the a different cell substrate:
      (i) Hepatitis A vaccine
      (ii) Chikungunya vaccine
      (iii) Measles
      (iv) Rubella
   (c) Drug substance for the following human vaccines can be manufactured with major changes in the manufacturing facility but with a different cell substrate:
      (i) Flu vaccine

13.7 Forward-Looking Statements

The facility has been set up in collaboration with DBT, with the strategic intent of having requisite vaccine infrastructure readily available during pandemic situations, in national interest. Following are the options available in the current scenario:

1. Continue with the objective of Covaxin drug substance manufacturing.
2. Change of objective to manufacture any other human vaccine.
3. Change of objective to manufacture any other veterinary vaccine.

At Hester, it is our endeavor to contribute toward Government of India's mission on Atmanirbhar Bharat in the field of human and veterinary healthcare (Figs. 13.1 and 13.2).

Fig. 13.1 Hester Biosciences, Kadi plant

Fig. 13.2 Facility photos of the BSL-3 lab

Fig. 13.2 (continued)

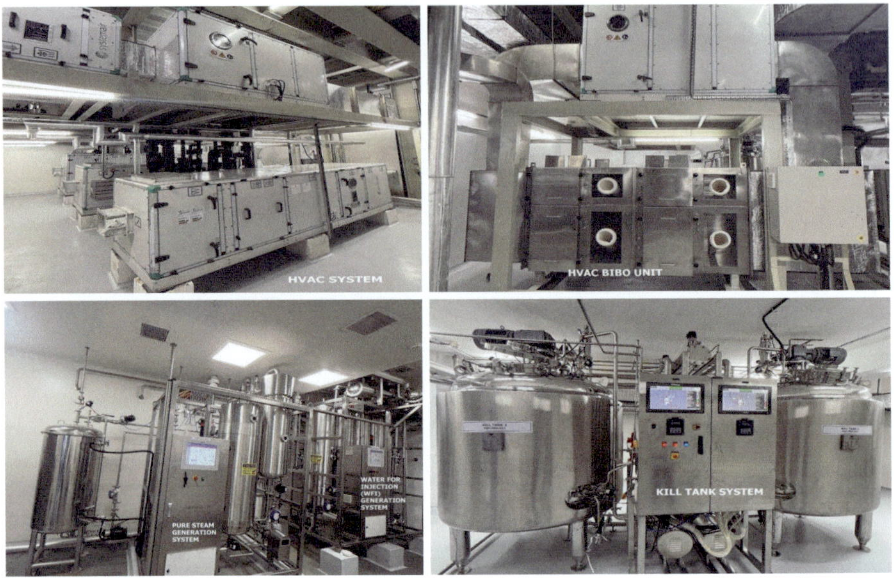

Fig. 13.2 (continued)

Reference

1. https://www.hester.in/

14 Make in India Vaccines: A True Story of Atmanirbhar Bharat—Indian Immunologicals Limited's Contribution to Mission Covid Suraksha

K. Anand Kumar and Mukul Gaur

14.1 Indian Immunologicals Ltd.—An Introduction

Indian Immunologicals Ltd. is a One Health company with 40 years of glorious existence. IIL is a research-to-market company offering cost-effective solutions to address various unmet needs in healthcare business. IIL manufactures and markets several veterinary biologicals and human vaccines in India and abroad. IIL is the largest producer of human Anti-Rabies vaccine in the world. IIL is one of the largest supplier of human vaccines to the Ministry of Health and Family Welfare and the largest supplier of animal vaccines to Ministry of Animal Husbandry, Dairying and Fisheries (Government of India). IIL operates four manufacturing plants and employs 2000 people. In 2022–23, IIL clocked a turnover of Rs.918 crores and PBT of Rs.252 crores.

Indian Immunologicals Limited (IIL) was established by National Dairy Development Board (NDDB) in 1982, with the objective of making vaccines available at an affordable price. Today, IIL holds a market leadership position in veterinary biologicals in India and operates one of the largest plants in the world for veterinary vaccines. IIL started manufacturing human vaccines in 1998, and it is the second company in the world and first in India to launch purified Vero cell rabies vaccine (PVRV). IIL also exports vaccines to 60 different countries.

IIL manufactures multiple vaccines for multiple species using various platform technologies. Livestock vaccines produced by IIL include vaccines for foot and mouth disease, theileriosis, foot and mouth disease (combination vaccine), hemorrhagic septicamia and black quarter, blue tongue, enterotoxaemia, Brucella (S19 and RB51), porcine cysticercosis, goat pox, and classical swine fever. IIL is the only producer of multivalent companion animal vaccines in India. On the companion

K. A. Kumar (✉) · M. Gaur
Indian Immunologicals Ltd., Hydrabad, Telangana, India
e-mail: anandkumar@indimmune.com

© The Editor(s) (if applicable) and The Author(s), under exclusive license to Springer Nature Singapore Pte Ltd. 2025
R. Swarup et al. (eds.), *At War with the Single Strand*,
https://doi.org/10.1007/978-981-96-8122-8_14

animal front, IIL produces vaccines for Rabies, Canine Parvo, Canine Distemper, Canine Parainfluenza, Canine adenovirus (CAV1 and CAV2), and Leptospirosis.

IIL has successfully developed several vaccines in the human health segment including recombinant Hepatitis B vaccine, Pentavalent vaccine, DPT vaccine, Td vaccine, Rabies vaccine, Hepatitis A vaccine, and Measles–Rubella vaccine and is currently working on Hexavalent vaccine, IPV trivalent Sabin-based vaccine, Bivalent Typhoid Conjugate vaccine, 15-valent Pneumococcal Conjugate vaccine, Dengue, Zika, and Codon-deoptimized live attenuated COVID-19 vaccine.

The vaccine platform for the manufacture of vaccines at IIL uses live attenuated virus, inactivated virus, recombinant subunit vaccines, live attenuated bacterial vaccines, bacterial toxoids, and conjugated polysaccharide vaccines. IIL manufactures bacterial and viral vaccines in liquid adjuvanted and lyophilized presentation. During the COVID-19 pandemic at the request from Niti Aayog, IIL rose to the occasion and augmented India's COVID-19 vaccine capacity by producing COVID-19 drug substance in record time. This effort has been lauded by Government of India. IIL has collaboration with several scientific and technological institutes and agencies like Harvard Medical School, USA, Centre for Disease Control (CDC), United States Army Medical Research Institute of Infectious Diseases, Global Alliance Livestock Vet Medicine (GALVmed), Pirbright Institute, UK, Griffith University, Australia, and NIAID, NIH USA, IISc, IVRI, etc.

IIL has embraced technology in many ways. IIL is the first company in the world to deliver animal vaccines by using drones. IIL is one the first company in India to introduce virtual medical representatives.

IIL as a part of corporate social responsibility has adopted government schools, provides fortified flavored milk to thousands of school children every day, and provides free vaccination to over a lakh of cows housed in various ghousalas. IIL has donated an Oxygen Generation plant to TIMS, Gachibowli and has contributed significantly to the PM and CM COVID-19 fund.

It all started in the last week of March 2021, when Government of India under the stewardship of Dr. Vinod K Paul, NITI Aayog reached out to Indian Immunologicals Limited (IIL), a wholly owned subsidiary of National Dairy Development Board (NDDB), for an initial discussion to come on board to augment the domestic manufacturing capacity of vaccine against SARS-CoV-2 as India was going through a widespread devastating impact of COVID-19 pandemic.

IIL has been a pioneer in the manufacturing of veterinary vaccines in India, and in four decades of existence, it has made foray in human vaccines and contributed significantly to Government of India's Universal Immunization Program (UIP). Human Biologicals Institute, a division of Indian Immunologicals Limited, has been engaged in the production of human vaccines since 1998 and has been in the forefront in phasing out unsafe nerve tissue rabies vaccine by being the first company in India to manufacture and supply cell culture-based human Rabies vaccine.

With this background, rising to the occasion in case of national emergency, IIL immediately agreed to the proposal to collaborate with Bharat Biotech International Limited (BBIL), a private entity and erstwhile competitor of IIL in the field of vaccines. Within 2 weeks of initial discussion, a Confidentiality Agreement (CDA) was

Fig. 14.1 IIL's manufacturing facility at Karkapatla village, District Siddipet, Telangana

signed between IIL and BBIL on March 13, 2021, followed by a Technical Collaboration Agreement (TCA) on March 20, 2021. As per the agreement, IIL was to manufacture drug substance which was to be supplied exclusively to BBIL for converting to drug product (Fig. 14.1).

In the ensuing period, several agreements were signed in quick succession between IIL and BBIL to facilitate the technology transfer. The key element of agreement between IIL and BBIL was that technical collaboration was limited only to the manufacture of drug substance by IIL, but the manufacture of drug product was not part of the agreement. Drug substance manufactured by IIL was to be supplied exclusively to BBIL.

One of the critical aspects was that all necessary regulatory approvals were required to be in place in an accelerated timeline. The drug regulatory authority of India fast-tracked the approvals without compromising the regulatory norms. IIL received the NOC for test license (CT10) applied on March 16 to DCGI and NOC (CT11) on March 26. The test license was applied (Form 30) with State DCA on March 30, and IIL received the test license in Form 29 on April 1, 2021. Subsequently, in discussion with BBIL, it was decided that BBIL will apply for loan licence using IIL's manufacturing facility. An audit was conducted by CDSCO and the state drug department on July 26 and 27, 2021. Based on the positive outcome of the audit of IIL's facility, BBIL obtained loan license in Form 28DA to manufacture whole virion, inactivated corona virus bulk at IIL's modified Rabies vaccine manufacturing facility at Karkapatla on August 12.

14.2 Technology Transfer and R & D Scale Trials

BBIL shared the brief details of the manufacturing process for the drug substance, process flow, MFR, DS specifications, qPCR testing method for virus quantification, and specifications of raw materials and consumables.

The virus seed was transferred from BBIL on April 16, 2021, after obtaining the test license and MTA was signed. IIL initiated the first trial at R & D-scale bioreactor (packed bed bioreactor), and the DS sample was submitted to Bharat Biotech on May 10 for analysis at their end. At this stage, there were two major challenges: IIL used its own cell banks and upstream technology that were different from BBIL; BBIL was using microcarrier-based technology in batch mode and IIL embarked with perfusion technology to fast-track the project, keeping in mind the existing infrastructure.

Weekly review meetings were conducted between BBIL and IIL to discuss the progress and technical issues in the initial phase of this project. Meanwhile, employees engaged in development, manufacturing, and testing were vaccinated in phases with the help of State and Central Government agencies.

14.3 Funding and Approvals

A total of ₹ 60 crores was granted for the project by BIRAC. While documentation and approval processes were on going on, IIL used its own funds to kick-start the project without waiting for government funding.

14.4 Facility-Readiness

In phase-I, the Rabies antigen facility at our Karkapatla site near Hyderabad was repurposed for the manufacture of COVAXIN drug substance. This required suspension of manufacturing of Rabies antigen at Karkapatla, and this was only feasible because IIL has a dedicated facility at Ooty, Tamil Nadu for the production of Rabies vaccine, which was operational in full capacity round the year even during the pandemic.

The Rabies facility at Karkapatla was quickly repurposed within 6 to 8 weeks; major changes made were providing once-pass-through HVAC for the virus culture area, BIBO for the isolator, installing shower toward the personnel exit from the live virus area, and re-arrangement of the equipment as per the COVAXIN drug substance process flow. Besides, the COVAXIN drug substance required a chromatography system for purification, which was not available with IIL. It was estimated to take approximately 8 to 12 weeks to import this system. IIL could manage to arrange a chromatography column and system within 6 weeks with some deft maneuver and consistent follow-up. This was followed by facility validation to meet cGMP requirements.

Cell propagation for the first engineering lot at large scale started in May, and the first virus infection in the bioreactor was carried out in June 2021.

As already explained that the major difference in process (how the cells and virus are grown) was in bioreactor technology, BBIL was using micro-carrier-based technology in batch mode, while IIL used packed bed bioreactor in perfusion mode. Optimization of virus culture in the perfusion bioreactor was a challenge due to the cytopathic nature of virus, as frequent sampling was not possible to collect for the observance of the cytopathic effect. This required quick standardization of the process of virus culture, collection of viral harvest, followed by inactivation, and purification of inactivated harvest by chromatography with the first large-scale lot itself as multiple trials could delay the production of drug substance and eventually delay

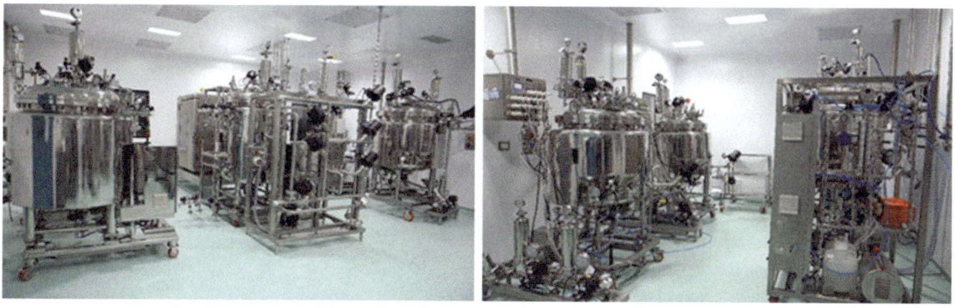

Fig. 14.2 Drug substance (viral antigen) manufacturing unit

the availability of drug substance for formulation at BBIL. The optimization of the process was achieved with the first lot itself, and the production of drug substance started in full swing, and this enabled us to transfer the first lot of drug substance to BBIL on August 13, 2021. Till March 2022, IIL produced approximately 22 million doses of drug substance and have supplied 20.15 million doses of drug substance.

14.5 New Viral Antigen Facility

A manufacturing facility for the production of viral vaccine antigen was planned at IIL's Karkapatla site in mid-2020; this was designed with a total built up area of 1950 m^2 and also had a BSL-3 manufacturing suite. The facility was designed to meet cGMP guidelines and biosafety regulation. Work on this new viral antigen facility started in November 2020. As IIL came on board for capacity augmentation of vaccine against SARS-CoV-2 under the mission Covid Suraksha, it was decided to equip and use this new facility for further increase of manufacturing capacity of COVAXIN drug substance. Despite several hurdles faced due to lockdown because of the pandemic, civil construction of the facility was completed in July 2021. There were several challenges due to worldwide lockdown leading to delay in the supply of critical process equipment; however, this facility was completed and qualified in October 2022 (Fig. 14.2).

IIL manufactured drug substance exclusively for BBIL, and the last drug substance lot was manufactured in March 2022, and further manufacturing was discontinued as BBIL no longer required the drug substance. The new viral antigen manufacturing facility built is currently being used for other viral vaccines of national importance.

The pandemic has taught us several lessons; a ready and available infrastructure and experienced manpower give us an advantage in terms of speed in any eventuality. As an important player in the field of vaccines of this vast country, we need to be in the state of pandemic preparedness all the time.

Competing Interest The authors have no conflicts of interest to declare that are relevant to the content of this chapter.

Science Behind COVID Vaccines

15

G. Padmanaban

Seven coronaviruses are of relevance to human diseases. 229 E, NL63, OC43, and HKU1 would have infected the human population with common cold, but SARS (Severe Acute Respiratory Syndrome), MERS (Middle-East Respiratory Syndrome), and SARS-2 have major implications for human health and mortality. These viral infections have visited, almost every decade, since the millennium, starting the year 2000. In particular, SARS-2 has been a major unprecedented affliction with over 768 million infections and 6.9 million deaths, reported to WHO. SARS-2 was first reported in Wuhan, China, in late December 2019. Vaccines have played a major role in limiting severe diseases and deaths, although the virus has been able to infect. COVID-19 vaccines are widely credited to have prevented an additional 14.14 and 19.8 million deaths during December 8, 2020 to December 8, 2021 in 185 countries [8].

Many countries prioritized vaccination of high-risk population such as the elderly with health issues and healthcare workers. Figure 15.1a,b shows the spike protein structure and the virus life cycle [1].

Part of the spike can extend and attach to a protein called ACE2 (in yellow below), which appears on particular cells in the human airway. The virus can then invade the cell.

The gene for the spike protein in SARS-CoV-2 has an insertion of 12 genetic letters: *ccucggcgggca*. This mutation may help the spikes bind tightly to human cells—a crucial step in its evolution from a virus that infected bats and other species. A number of scientific teams designed vaccines that could prevent the spikes from attaching to human cells.

G. Padmanaban (✉)
Department of Biochemistry, Indian Institute of Science, Bangalore, India
e-mail: geepee@alumni.iisc.ac.in

© The Editor(s) (if applicable) and The Author(s), under exclusive license to Springer Nature Singapore Pte Ltd. 2025
R. Swarup et al. (eds.), *At War with the Single Strand*,
https://doi.org/10.1007/978-981-96-8122-8_15

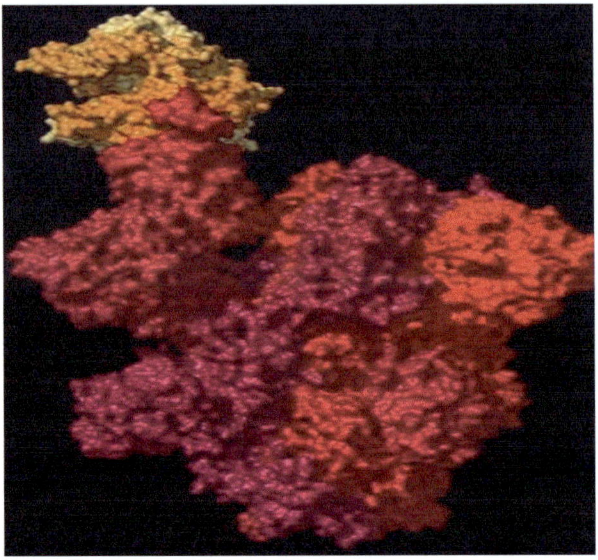

Fig. 15.1 Spike protein trimeric structure

15.1 Vaccine Platforms [2, 3]

The vaccine platforms are being employed for SARS-CoV-2. Whole virus vaccines include both attenuated and inactivated forms of the virus. Protein and peptide subunit vaccines are usually combined with an adjuvant in order to enhance immunogenicity. The main emphasis in SARS-CoV-2 vaccine development has been using the whole spike protein in its trimeric form, or components of it, such as the RBD region. Multiple nonreplicating viral vector vaccines have been developed, particularly focused on adenovirus, while there has been less emphasis on the replicating viral vector constructs.

India, China, Russia, and a few other countries introduced the classical inactivated viral vaccine. For example, the Indian vaccine, Covaxin, introduced by Bharat Biotech International Ltd., includes whole-virion-inactivated SARS-CoV-2 antigen using Beta-propiolactone, TLR7/8 agonist (imidazoquinolinone class molecule), 2-phenoxyethanol (anti-infective), and phosphate buffer. The virus was isolated from an Italian tourist at National Institute of Virology, Pune (strain: NIV-2020-770), grown in large quantities using Vero cells and then inactivated. The inactivated virus preparation was then mixed with an aluminum-based adjuvant, Alhydroxyquim. The vaccine is injected in two doses through the intramuscular route and seems to elicit humoral and cellular responses with a bias to TH1 immune response. The vaccine received marketing authorization on January 3, 2021 and WHO authorization for emergency use on November 3, 2021. This vaccine accounts for less than 15% of vaccine doses administered in India. Subsequently, BBIL developed a nasal vaccine version with DBT support, mentioned later. But, China vaccinated bulk of its population with its own versions of whole viral-inactivated vaccines.

As mentioned earlier, other than the whole inactivated virus, rest of the global vaccines have targeted the spike protein or its receptor-binding domain (RBD) using different strategies.

15.2 Adenovirus Vector Vaccines [4]

These contain adenovirus DNA, encoding the spike protein gene, but nonreplicating. The viral vector-based vaccines against COVID-19 express the protein antigen, eliciting an immune response. Examples of this type of vaccines are the Oxford–Astra Zeneca COVID-19 vaccine, the Sputnik V COVID-19 vaccine, Convidecia, and the Janssen COVID-19 vaccine. Convidecia and Janssen are both one-shot vaccines and so easier to use in the field. It can be stored under ordinary refrigeration conditions for several months. Sputnik Virus Ad26's first dose is the same as Janssen's only dose, and Ad5's second dose is the same as Convidecia's only dose. In view of the Delta spread in August 2021, developers of Sputnik V suggested Ad26 be used as a booster to the mRNA vaccine.

15.3 ChAdOx1n CoV-19 (AstraZeneca)

ChAdOx1 is the replication-deficient adenovirus vector version of ChAdY25, first isolated from chimpanzee fecal samples, since there was a concern that the population may carry pre-immune antibodies to the adenovirus. The ChAdOx1 nCoV-19 vaccine encodes the tPA leader sequence fused to the 5′ end of a codon-optimized version of the full-length SARS-CoV-2 spike sequence. Mice and non-human primate studies indicated high immunogenicity for neutralizing live virus and antigen-specific IFNγ-producing CD8+ and CD4+ T-cells. In addition to mice and NHPs, the ChAdOx1nCoV-19 vaccine was also tested in ferrets. Serum Institute of India, Pune, has turned out to be the major power house to manufacture the AstraZeneca vaccine for India and the globe. It has also diversified to make vaccines on other platforms as well.

Bharat Biotech International Ltd. was the first to introduce a COVID nasal vaccine. It is a *recombinant replication-deficient adenovirus-vectored vaccine* (Incovacc) with a prefusion-stabilized spike protein. It triggers an immune response in the mucosal membrane. It is held to be effective against Omicron variants that replicate in the upper respiratory tract before entering the lungs.

15.4 Subunit Vaccines [5]

Subunit vaccines in general contain recombinant protein antigen, expressed using an appropriate vector, made in large quantities and purified. Examples of this type are Novavax COVID-19 vaccine, the peptide vaccine EpiVac Corona, ZF2001, MVC-COV1901, Corbevax, the Sanofi–GSK vaccine, and Soberana02 (a conjugate vaccine).

The V451 vaccine was in clinical trials that were terminated after it was found that the vaccine may potentially cause incorrect results for subsequent HIV testing.

15.4.1 Corbevax-Biological E Pvt., Ltd. (Example)

RBD (receptor-binding domain) was expressed in *P pastoris* adsorbed to the adjuvant Alhydrogel (Alum), in combination with another approved adjuvant, CpG1018. The initial construct and production process were developed at Texas Children's Hospital, Houston. The heterologous booster dose for Covishield and Covaxin was approved by DCGI, Union Health Ministry. Bivalent vaccine involved Corbevax booster after Covaxin or Covishield vaccination.

15.4.2 Nucleic Acid Vaccines

The plasmid DNA encoding the protein antigen is injected intramuscularly. The expressed protein generates an immune response. The advantage of the DNA vaccine is its stability. But, the DNA has to enter the cells and then cross the nuclear membrane to be transcribed into the DNA vaccine [6] mRNA, which is translated in the cytoplasm to the corresponding protein.

Zycovid-D is the first human DNA vaccine introduced by Zydus Cadila, a pharma major in India. The plasmid encodes the spike protein and also carries unmethylated CpG motifs to enhance immune response. The plasmid is expressed in *E coli* cells. A unique aspect is the needleless delivery using intradermal injection using a spring-powered jet injector.

15.4.3 Messenger RNA Vaccine

The plasmid DNA strands are separated, and then RNA polymerase is used to transcribe the DNA to mRNA. The mRNA has to undergo capping at the 5′ end and then injected. The mRNA is translated in situ, and the protein antigen elicits an immune response. Self-amplifying mRNA involves additional factors to form RNA-dependent RNA polymerase complex that leads to the amplification of mRNA (Fig. 15.2).

Gennova has made two mRNA vaccine products. GEMCOVAC™-19, thermostable, and GAMCOVAC-OM, thermostable and Omicron-specific booster vaccine. The vaccine is delivered intradermally using a needle-free injection system mentioned earlier. The mRNA is adsorbed on the surface of the nano-lipid emulsion unlike it being entrapped as in other mRNA vaccines. mRNA vaccines have the advantage that they need not enter the cell nucleus to be transcribed but translated straight away in the cytoplasm. Gennova has carried out a successful clinical trial with the mRNA preparation against the Omicron variant [7].

Nucleic acid vaccines have the flexibility to change the base sequence, and therefore a new vaccine can be prepared quickly against a variant.

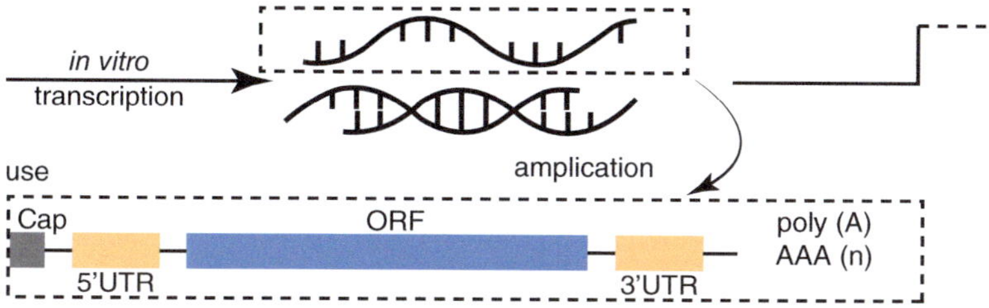

Fig. 15.2 mRNA preparation in vitro

15.5 Vision Ahead

This article deals with principles behind the COVID vaccines made in India. The country came forward to help other countries with millions of vaccine doses and has developed the technology with support from partners abroad. DBT, ICMR, Health Ministry, Scientific Institutions, and industries have played a major role in all the developments. WHO has officially announced the end of the pandemic, but the tiny virus can pose new challenges and become endemic. The country is ready to handle variants and even new viruses that can cause a pandemic in future. But, technology keeps changing to make and provide new vaccines within a couple of years, unlike the decade timeline experienced in the past. An interesting development is the importance gained by vaccines as a major tool for disease protection, although they have indeed saved millions over the millennia. Figure 15.2 gives a gist of various vaccine platforms, and different countries have invested in the development of different vaccine designs. It is also important for the government, regulatory authorities, and scientific establishments to build a new India that can really enter into disease eradication phase and achieve the relevant sustainable development goals set by the United Nations.

Acknowledgments Vaccines in India were developed with the funding support from the Department of Biotechnology, Government of India.

Competing Interest The authors have no conflicts of interest to declare that are relevant to the content of this chapter.

References

1. Zhang J, Xiao T, Cai Y, Chen B. Structure of SARS-CoV-2 spike protein. Curr Opin Virol. 2021;50:173–82.
2. Applied Biological Materials Inc. SARS-CoV-2 infection and replication cycle. Brochure.
3. Nagy A, Alhatlani B. An overview of current COVID-19 vaccine platforms. Comput Struct Biotechnol J. 2021;19:2508–17.
4. Chavda VP, Bezbaruah R, Valu D, Patel B, Kumar A, Prasad S, Kakoti BB, Kaushik A, Jesawadawala M. Adenoviral vector-based vaccine platform for COVID-19. Current Status Vaccines. 2023;11:432.

5. Heidary M, Kaviar VH, Shirani M, Ghanavati R, Motahar M, Sholeh M, Ghahramanpour H, Khoshnood. A comprehensive review of the protein subunit vaccines against COVID-19. Front Microbiol. 2022;13:927306.
6. Khalid K, Poh CL. The development of DNA vaccines against SARS-CoV-2. Adv Med Sci. 2023;68:213–26.
7. Saraf A, Gurjar R, Kaviraj S, et al. An omicron-specific, self-amplifying mRNA booster vaccine for COVID-19: a phase 2/3 randomized trial. Nat Med. 2024;30:1363–72.
8. Watson OJ, Barnsley G, Toor J, Hogan AB, Winskill P, Ghani AC. Global impact of the first year of COVID-19 vaccination: a mathematical modelling study. Lancet Infect Dis. 2022;22(9):1293–302. https://doi.org/10.1016/s1473-3099(22)00320-6. PMC9225255. PMID35753318

Establishing and Leveraging Clinical Trial Networks and Health Surveillance Systems: Crucial Strategies During the COVID Pandemic

16

Narendra Kumar Arora and Shikha Taneja Malik

16.1 The Need

Department of Biotechnology (DBT), Ministry of Science and Technology, launched the National Biopharma Mission (NBM) or Innovate in India for Inclusiveness (i3) in 2017. The program was co-funded by a World Bank loan and the Government of India and was implemented through Biotechnology Industry Research Assistance Council (BIRAC). The focus of the program was to bolster national product development capabilities in vaccines, biotherapeutics (including monoclonal antibodies), and medical devices to a globally competitive level over the next decades. Additionally, the imperative of strengthening the ecosystem to enhance clinical research capabilities, particularly competence, to undertake nonregulatory and regulatory compliant trials was also considered as a priority investment domain.

16.2 The Pandemic

The World Health Organization (WHO) declared the outbreak of COVID-19 as a Public Health Emergency of International Concern (PHEIC) on January 30, 2020, which further was declared as a pandemic on March 11, 2020. After it was declared a pandemic due to the alarming levels of its spread and severity, a deemed necessity for accelerating the development of diagnostics, vaccines, novel therapeutics, and re-purposing of drugs for this novel coronavirus arose.

N. K. Arora (✉)
The INCLEN Trust International, New Delhi, India
e-mail: nkarora@inclentrust.org

S. T. Malik
Scientific Affairs, Drugs for Neglected Diseases initiative (DNDi), NBM, BIRAC, New Delhi, India

© The Editor(s) (if applicable) and The Author(s), under exclusive license to Springer Nature Singapore Pte Ltd. 2025

R. Swarup et al. (eds.), *At War with the Single Strand*,
https://doi.org/10.1007/978-981-96-8122-8_16

BIRAC and DBT swiftly took action to bolster the preparedness, readiness, and response to the COVID-19 pandemic. In early 2020, two "Request for Proposal (RFP)" for the DBT-BIRAC COVID-19 Research Consortium were launched. These RFPs invited project proposals aimed at developing diagnostics, vaccines, novel therapeutics, drug repurposing, or any other intervention to combat COVID-19 with participation from both industry and academia.

16.3 Harnessing Established and New Capacities from the NBM and DBT Programs to Combat the Pandemic

In response to the COVID-19 pandemic, the Department of Biotechnology (DBT), Ministry of Science and Technology, had been working relentlessly to address the requirement of a COVID vaccine with high priority. DBT through Biotechnology Industry Research Assistance Council (BIRAC)/National Biotechnology Mission (NBM) aimed to strengthen the facilities and capabilities to streamline and accelerate vaccine development in the country and be capable of effectively implementing Mission COVID Suraksha, through the establishment of a Mission Implementation Unit (MIU).

The anticipatory efforts initiated through the National Biopharma Mission Program in 2017–2018 facilitated the rapid enhancement and utilization of critical resources and facilities for the development of COVID-19 vaccines within the country.

Since its inception, the impact of the NBM Program resulted in strengthening 21 research facilities, establishment of five [5] translational research consortia, 11 clinical trial sites with access to more than 800,000 healthy population, and establishment of disease registries for four diseases at another 36 hospital sites. Besides these, efforts had been taken to establish indigenous capacities for the manufacturing of raw materials and components being utilized for vaccine and antibody manufacturing [1]. The mission has also made a huge impact on the medical device and diagnostic sector in the country.

Any vaccine development takes an average of 10–15 years, but considering the dire necessity for COVID-19 vaccine to be accessible to the citizens, DBT established Mission COVID Suraksha, with the mandate, to integrate and to streamline available resources toward a warpath to fast-track vaccine development. This being a national mission aimed at bringing a safe, effective, convenient, and available COVID vaccine to the citizens of the country as soon as possible, with a focus on "*Atmanirbhar*", and keeping our responsibility to serve not just our country but the entire world, with a focus on "*Vasudhaiva Kutumbakam*". During the short period of 4 years, with the financial and technical support marshalled by NBM, BIRAC, and DBT, the following five COVID-19 vaccines were developed as part of Mission COVID Suraksha and also received emergency marketing authorization by the Drug Controller General of India: ZyCOV-D®, CORBEVAX®, GEMCOVAC-19®, GEMCOVAC-OM®, and iNCOVACC®.

NBM or "Innovate in India for Inclusiveness" (i3), started off to identify the barriers to conduct clinical research/trials in India, regulatory as well as nonregulatory. Multiple brainstorming sessions, stakeholder meetings, and one-on-one discussion with experts had been conducted with the following overall aim:

- To understand the strengths and weaknesses of different sectors, i.e., hospital- and community-based trials, industry- and academia-led trials, treatment and prevention trials, and research studies in community settings.
- Outline possible solutions, requirements toward setting up clinical trial or surveillance networks, and their model of functionality.
- Seek recommendations to enable NBM develop appropriate strategies and "call for proposals" to address these challenges/gaps.

Several barriers to conduct clinical research or clinical trials were highlighted by the experts:

1. Infrastructure
2. Availability of finances
3. Human capacity
4. Ethical and regulatory system-related obstacles
5. Research-supportive environment
6. Operational and administrative barriers

Planning of field-based trials requires reliable estimates of disease burden and input of accurate incidence data from different geographical zones. For conduct of such trials, it is very important to have a detailed demography of the healthy population. Further, a well-established clinical trial site with necessary infrastructure, human capacity, quality management systems, including access to NABL-accredited laboratory facilities and disease epidemiology data are critical pre-requisites for the conduct of Good Clinical Practice (GCP)-compliant trials.

The National Biopharma Mission (NBM) committed significant time to conduct brainstorming sessions with stakeholders and interacted with key opinion leaders from across the country, including Dr. Sanjay Mehendale, Director Research, Hinduja Hospital and Medical Research Centre, Mumbai, and Dr. CS Pramesh, Director, Tata Memorial Hospital, Mumbai, and received global inputs through experts participating in DBT an BIRAC committees, e.g., NIAID-NIH experts who were part of the Indo-US Vaccine Action Program and BMGF's national and global teams. This helped NBM align the Request for Proposals (RFPs) with international standards and also incorporated the insights gained from NIH's Clinical Trial Units (CTUs).

In 2018, NBM initiated supporting indigenous development of new and novel vaccine candidates for Dengue and Chikungunya viruses. In addition to candidates supported by the mission, there were several other vaccine candidates under development in the country for these two infections. As background for the forthcoming trials for these vaccines, valid and reliable data were considered essential from different geographical zones of the country for determining the infection and disease attack rates, hospitalizations, severe disease, deaths, medical visits, and costs. Keeping in view the above requirements, it was important to have field sites that regularly provided disease burden data, had access to all age groups of target population, and ability to do long-term follow-up of each enrolled subject, to do cluster-based randomized trials when required, to map private and public health services, so as not to miss hospitalized cases, and have access to accredited laboratories.

16.3.1 The First Step

With this in view, NBM launched four requests for proposals (RFPs) on August 15, 2019 for strengthening community-based vaccine trial capacities in healthy population cohorts and establishment of clinical trial network for hospital-based trials of diabetes, ophthalmology, rheumatology, and oncology. The focus of this chapter is on the community-based field sites that played a key role in the pandemic, contributing to COVID vaccine development and surveillance studies.

16.3.2 Request for Proposals (RFPs) Published

1. Establish new Demographic Surveillance Sites (DSSs)/Demographic and Health Sites (DHSs)/Demographic, Development, and Environmental Surveillance Sites (DDESS) within the country maintaining geographical representation and to study the epidemiology of Dengue and Chikungunya in different age groups at these sites.
2. To study epidemiology of Dengue and Chikungunya in different age groups at the existing DSS/DHS/DDESS and to further prepare the sites for conduct of GCP-compliant field-based clinical trials.
3. To establish data management platform: an IT platform for community-based data collection, analysis, and reporting.

16.3.3 Outreach

The government launched a new initiative to establish population sites, with the aim of enhancing national surveillance and clinical trial capacity. The NBM sought to communicate with all potential grantees, explaining the rationale, necessity of the initiative, and its vision. Despite numerous unresolved challenges, the NBM remained open to suggestions and collaborated with investigators active in the field, within communities, and in cohort establishment.

Nationwide webinars were conducted by the NBM to address queries, and contact details were shared for further clarification needed during the application process. Initially, applications were invited in the form of a brief

Letter of Intent (LoI). Following an evaluation by the Area Review Panel (ARP), shortlisted Principal Investigators (PIs) were invited to submit full proposals.

During the proposal submission period, site visits were conducted by the officials from the NBM and the Clinical Development Services Agency (CDSA), their clinical consultancy partner. These visits were important to assess the existing capacities at the field offices or sites. Throughout the process, the PIs were guided to ensure they fully understood the initiative and had an equal opportunity to compile a robust full proposal.

After a multi-tier review of BIRAC, legal, financial, technical (multiple committees: ARP [area review panel], SAG [scientific advisory group], and TAG [technical advisory group]), a total of 11 grantees were selected for establishing the community-based surveillance network. SAG-Clinical Trial Network (CTN) was constituted with the representation of national experts. These included senior technical officials from the science departments, public and private hospitals, former industry personnel and academicians with pharmacology, or public health expertise. The TAG Chairperson Professor KS Reddy, President of Public Health Foundation of India (PHFI), with his immense experience in public health played a key role in shaping and guiding the selected grantees under CTN.

16.4 Funding Mechanism

NBM, BIRAC embarked on a novel initiative. The financial backing needed for these sites differed significantly from that of other NBM grantees. To ensure appropriate support for strengthening community work, Standard Operating Procedures (SOPs) tailored for the network were formulated and implemented.

Community work presented its distinctive set of challenges. For instance, in patriarchal regions where women were constrained to interact with outsiders, male field staff to facilitate communication with male family members were also hired. Combined female and male teams worked very well. Our stakeholder interactions revealed other unique human and cultural issues at various sites, landscape challenges, and deep sense of suspicion of the data collection work by the communities and necessity of sustained social mobilization as confidence-building measure and fostering long-term relationships.

Field staff serves as the backbone of public health, playing a crucial role in bridging the gap between communities and healthcare systems, and their efforts are instrumental in implementing health interventions and monitoring their impact at the ground level. Expenditures such as accidental insurance, rentals (including guest house and server rentals), and utilities like air conditioning, telephone bills, and electricity bills were permitted for the functioning of the field office and to ensure the welfare of the field staff.

16.5 Project Risk Management Plan

To comply with DBT, World Bank, and NBM, BIRAC's mission to promote innovation and self-sufficiency in the biotechnology sector while striving to reduce any social and environmental risks in its activities, fund recipients were guided to identify the risks related to project, environment (including occupational health and community), and conduct of clinical studies.

The sites were also strengthened and monitored on two important components: Clinical Research and Validation Management Framework (CRVMF) [2], and Environment, Occupational Health, and Safety Management Framework (EMF) [3].

CRVMF ensured adherence to bioethics principles in the conduct of clinical study. The plan abides by the legal, regulatory, and ethics requirements as per the national and global guidelines and the social safeguard policies.

EMF was implemented for grantees to endeavor for minimizing/mitigating adverse environmental and social impacts/risks and to ensure that such minimization or mitigation of impacts and risks meets the requirements of laws and regulations of Government of India (GOI) and states, and responsive toward environmental and social safeguard requirements of BIRAC.

16.5.1 DBT'S Resource of Indian Vaccine Epidemiology Network (DRIVEN)

In the spirit of BIRAC's inventive acronyms, DRIVEN was a unique name for a distinctive initiative. DRIVEN was established in 2020 and has been actively engaged in enhancing capacities and conducting surveillance studies and clinical trials (Fig. 16.1). DRIVEN network currently covers a total of 11 community sites

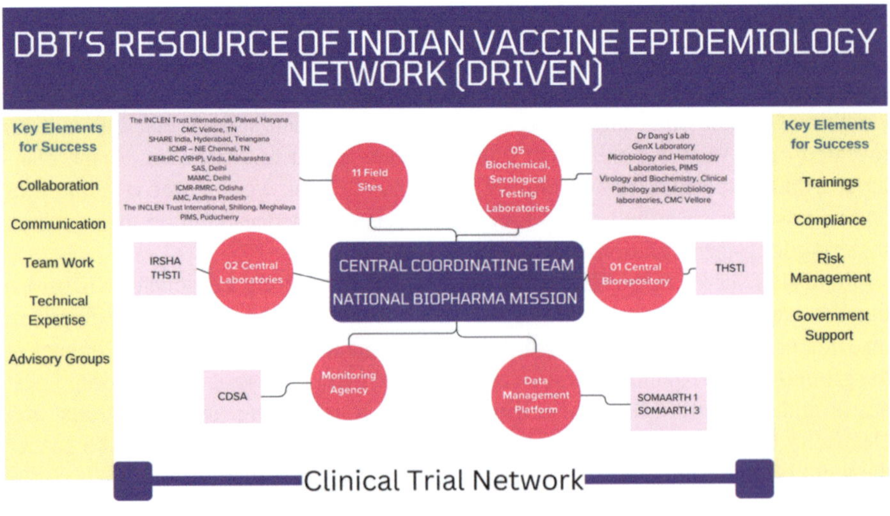

Fig. 16.1 Components and stakeholders

located in Delhi, Haryana, Odisha, Meghalaya, Puducherry, Tamil Nadu, Maharashtra, Telangana, and Andhra Pradesh, with access to rural, semiurban, urban, and tribal populations.

16.6 Establishment of Six New Demographic, Development, and Environmental Surveillance Sites (DDESS)

Establishment of six new DDESS in different geographical locations across India was supported by the NBM for collecting demographic data electronically for future studies through a common protocol with the objective to establish a population-based geographical cohort toward having Demographic and Health Surveillance (DHS) sites. The size of each cohort was to be equivalent to or higher than 50,000 population. The focus was to have adequate human resource and training of personnel for conduct of GCP-compliant trials, engage the community in advance, and collect information on co-morbidities for preparing the sites that meet internationally accepted clinical and ethical standards.

The INCLEN Trust International, with their experience in developing protocols for establishing DDESS and their expertise of the electronic data collection platform for these sites, harmonized the processes to be followed to bring in uniformity in the methodology and steps to be followed in carrying out the study that eventually reflected in the quality of data collected and preparedness of all these sites for future clinical trials.

These six sites are as follows:

- CHRD Society for Applied Studies, New Delhi
- Maulana Azad Medical College, New Delhi
- Andhra Medical College, Vishakhapatnam, Andhra Pradesh
- ICMR-Regional Medical Research Centre, Bhubaneswar, Odisha
- Puducherry Institute of Medical Sciences, Puducherry
- The INCLEN Trust International, Mawphlang (East Khasi Hills, Shillong), Meghalaya

16.7 Establishing Serial Sero-Surveillance to Monitor the Trend of SARS-CoV-2, Dengue, and Chikungunya Infection Transmission in the General Population, India

Constitution and the presence of DRIVEN were considered an opportunity for expanding the network's scope and collaborating with additional partners, although this demanded significant additional resources (both in terms of time and finances). Pre-existing four field sites located in different geographies across India were supported by NBM for conducting acute febrile illness surveillance of Dengue and Chikungunya through a common protocol.

Amidst the pandemic, the scope of these four sites was further expanded to include COVID-19, and an additional site (KEMHRC-Vadu, Pune [Maharashtra]) was incorporated into the DRIVEN initiative. KEMHRC was being supported through the Grand Challenges India program.

All these five (05) sites had access to the existing cohort of around 250,000+ healthy subjects. These sites are as follows:

- The INCLEN Trust International, Palwal, Haryana
- Christian Medical College, Vellore, Tamil Nadu
- SHARE India, Medchal, Hyderabad, Telangana
- ICMR-National Institute of Epidemiology, Chennai, Tamil Nadu
- KEMHRC (VRHP), Vadu, Pune, Maharashtra

16.8 Central Labs for Immunogenicity Evaluation

The National Immunogenicity Biologics Evaluation Center (NIBEC), situated at the Interactive Research School for Health Affairs (IRSHA) in Bharati Vidyapeeth University, Pune, along with the Bioassay Laboratory at THSTI, Faridabad, was supported by the NBM and DBT. These facilities served as GCLP (Good Clinical Laboratory Practice) centers for conducting viral neutralization assays and also performed as the central laboratories. Both laboratories held NABL accreditation and had validated assays. To ensure robust data analysis, interlab comparisons (ILCs) were conducted between these two laboratories for the relevant assays.

16.9 NABL-Accredited Laboratories for Biochemical, Molecular, and Serological Testing

- Dr. Dang's Lab, Delhi
- GenX Laboratory, Odisha
- Microbiology and Hematology Laboratories, PIMS, Puducherry
- Virology and Biochemistry, Clinical Pathology, and Microbiology Laboratories, CMC Vellore, Tamil Nadu

Central Biorepository THSTI, Faridabad, was supported and entrusted to hold the biological samples collected from the DRIVEN network for different studies. The samples would be available for scientists from the DRIVEN network and the other institutions for studies in future.

16.10 Electronic Data Management (SOMAARTH Platform)

A strong recommendation by the stakeholders was to ensure secure and paperless data collection and its management as per the national guidelines. To address the same, proposals were requested to focus on the development of e-data management platform customized for HDSS/DDESS sites and the conduct of COVID, Dengue, and Chikungunya studies. The INCLEN Trust International, New Delhi, had developed an electronic data management platform, namely SOMAARTH-1 and SOMAARTH-3 for collecting data, reporting, analysis, and archival of community-based demographic, development, and environmental surveillance sites and associated studies.

SOMAARTH Data Platform captured the social, cultural, and geographic diversity of the DRIVEN sites. These sites by their design itself were well equipped with systems for community engagement, census enumeration, geospatial data, and environmental monitoring. DRIVEN sites are good clinical practices (GCP)- and good clinical laboratory practices (GCLP)-compliant. All the sites used a harmonized data framework covering demography, development, health, and environmental aspects. The electronic data collection platform facilitated rapid data sharing and validation process. The harmonized data framework was adapted for Indian contexts, i.e., geographical, socio-cultural settings. Demographic and environmental features were geo-mapped at these sites. The detailed understanding of traditional, cultural governance systems and community dynamics came very handy during the implementation of trials and other health intervention projects. The key datasets available at the SOMAARTH platform covered the size and composition of the population, density, household and neighborhood characterization, chronic diseases, WaSH practices, health-seeking behaviors, and community burial practices.

Linking health and demographic data with geographic information system (GIS) mapping will be used to improve planning for public health interventions, for example, to assess the correlation of malaria prevalence and anemia in the population, to analyze the effects of environment on early childhood mortality, and many more such important studies.

HDSS/DHS/DDESS will play an important role not only for the scientists planning clinical research studies but also for policy-making by the government, provide scientific- and evidence-based feedback on the impact of national programs, and defining priorities and strategies for improving population health.

Monitoring Agency Clinical Development Services Agency (CDSA), THSTI.

CDSA is an extramural unit of Translational Health Science and Technology Institute (THSTI) and through a national-level selection process was selected as a consultancy firm. They played key role in training the staff on GCP as per the need of the study sites. They also ensured that all systems followed during the study conduct are adequate which include but not limited to ethics clearance, compliance with ethics approval terms, and quality management assurance. They supported NBM in evaluation of the sites before initiation and during the studies.

Central Coordinating Agency It was National Biopharma Mission, BIRAC. The NBM team played a pivotal role in offering technical and administrative assistance to all partners from the very beginning, ensuring the successful execution of the initiative.

16.11 Salient Features of DRIVEN Network (Fig. 16.2)

- Population cohort: Access to more than 800,000 individuals at 11 sites.
- Six DDESS sites are GIS-enabled with built structures and environmental parameters mapped along with health and demographic information.
- Implementation of common protocol for studies.
- Implementation of harmonized SOPs.
- GCP-trained staff.
- Quality management systems.
- Linkage with NABL-accredited laboratories.
- All sites have Ethics Committees, and approval is obtained for every study undertaken at these sites.
- Central laboratories for validated assays.
- E-data management ensuring the accuracy, integrity, and security of the collected data. It also facilitates efficient data retrieval and analysis, thereby aiding in informed decision-making and compliance with global standards.
- In the DRIVEN network, the staff members are proficiently trained in Good Clinical Practice (GCP), Good Documentation Practice (GDP), and Good Clinical Laboratory Practice (GCLP). Leveraging the robust infrastructure, skilled workforce, and efficient systems, they have successfully competed for national and international research funding, furthering their commitment to ongoing studies. Additionally, some network partners proactively secured additional funding for further investigations at select sites.

The clinical research infrastructure established under the various missions like NBM, Indian Component of Coalition of Epidemic Preparedness (InDCEPI), and Grand Challenges India (GCI) was leveraged during the COVID-19 pandemic. With the learnings of establishing the DRIVEN network, additional 19 clinical trial sites were strengthened under the Mission COVID Suraksha to support the COVID-19 vaccine product development; 13 of these sites also participated in COVID vaccine trials. The country now has a network of 54 clinical trial sites with the established infrastructure, capacity, and skilled manpower to conduct clinical studies.

Numerous peer-reviewed international publications have emerged from the network partners, spanning both site-specific and multi-site research [4, 5]. Recently,

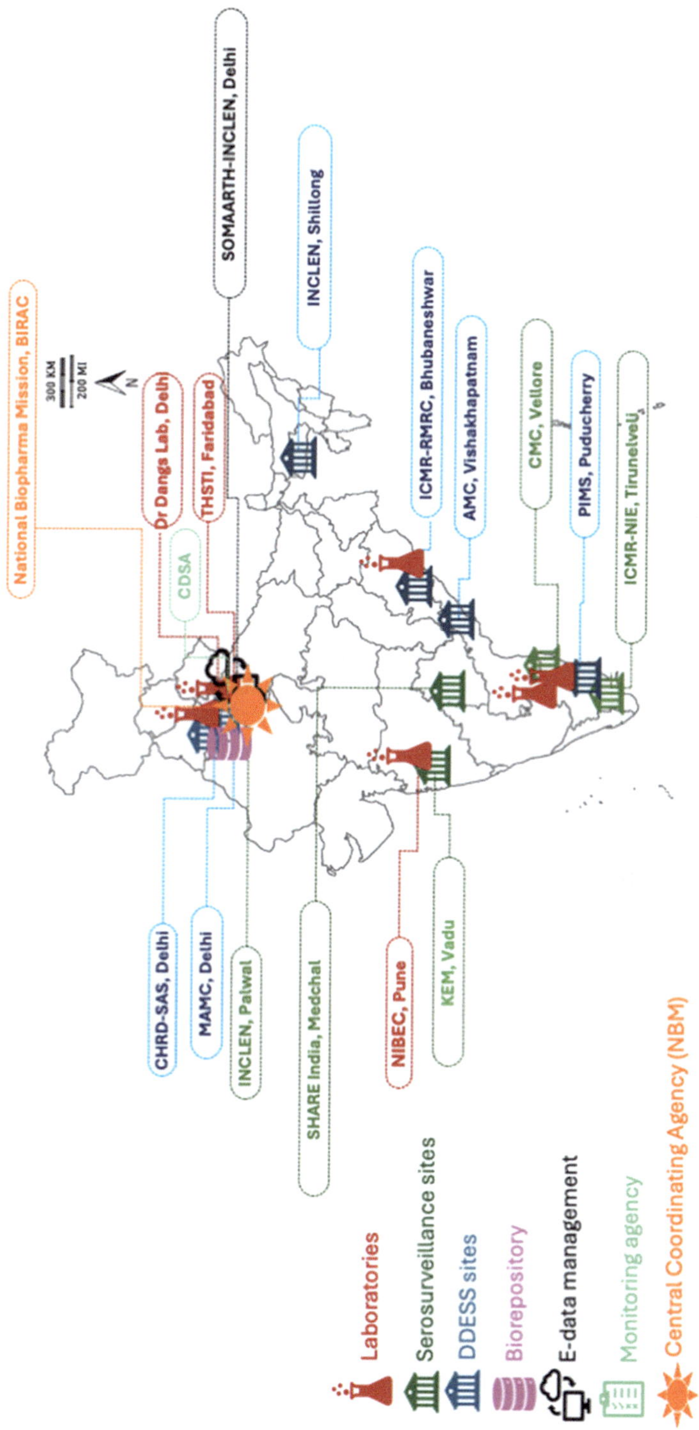

Fig. 16.2 DBT's Resource of Indian Vaccine Epidemiology Network (DRIVEN)

DRIVEN launched a monograph on setting up DDESS (Demographic, Development, and Environmental Surveillance Sites). Demographic, health, and environmental features are geo-mapped at these sites. This monograph collated the experience of different DRIVEN network partners for the setting up of a DDESS. This comprehensive compilation covers the entire spectrum of DDESS, from their design and establishment to the generation of actionable findings. This also shares the methodologies and various challenges faced during site development, including GIS mapping, ethical considerations, data collection methodologies, and ensuring the engagement and participation of local communities [6].

DRIVEN network and the CHOORD (Consortia of Hospitals in the areas of Oncology, Ophthalmology, Rheumatology, and Diabetology) network were leveraged to undertake pandemic-related vaccine and drug trials. Seven sites were leveraged for conduct of clinical trials of COVID-19 vaccine (Table 16.1).

16.11.1 COVID-19, Dengue, and Chikungunya Sero-Surveillance in a Cohort of 25,000

The more experienced five (05) DRIVEN sites were selected by NBM for conducting Dengue and Chikungunya incidence studies; additional support under NBM was pivoted to conduct sero-surveillance to monitor the trend of SARS-CoV-2 infection transmission in India in a multi-site, community-based longitudinal cohort study. The study involved follow-up of a cohort of 25,000 individuals (@ 5000 individuals per site) aged 2 years and above recruited from five existing demographic surveillance sites (DRIVEN) in India. The study protocol developed has been published in Frontiers in Public Health (2022) [4]. The data collection has been now completed, and analysis is ongoing. Such studies shall help to determine the burden of the disease at the community level in urban and rural Indian populations and guide in monitoring the trends in the transmission of the infection

Table 16.1 DRIVEN sites strengthened/established through DBT missions, leveraged for COVID-19 vaccine trials

Site name	COVID-19 vaccine
Centre for Health Research and Development—Society for Applied Studies (SAS), Delhi	COVOVAX, GEMCOVAC®, Sputnik
KEM Hospital Research Centre, Pune, Maharashtra	GEMCOVAC®
Medanta Institute of Education and Research, Gurgaon, Haryana	GEMCOVAC®, Corbevax™
Amrita Institute of Medical Sciences and Research Centre (AIMS), Kochi	Novavax, COVOVAX
JIPMER, Puducherry	r-BCG; COVOVAX
Christian Medical College Vellore Association, Vellore, Tamil Nadu	Corbevax™
INCLEN Trust International Palwal	*Sputnik V Vaccine, iNCOVACC,* GEMCOVAC®

Table 16.2 Names of the DRIVEN partners participating in the AFI study

1. The INCLEN Trust International, Palwal, Haryana	2. Christian Medical College, Vellore, Tamil Nadu
3. SHARE India, Medchal, Hyderabad, Telangana	4. ICMR—National Institute of epidemiology, Chennai, Tamil Nadu
5. CHRD Society for Applied Studies, New Delhi	6. Maulana Azad medical college, New Delhi
7. Andhra medical college, Vishakhapatnam, Andhra Pradesh	8. ICMR- regional medical research Centre, Bhubaneswar, Odisha
9. Puducherry Institute of Medical Sciences, Puducherry	10. The INCLEN trust international, Mawphlang (east Khasi Hills, Shillong), Meghalaya

(particularly SARS-CoV-2). Such surveillance also highlighted the risk factors for infection and could inform future control strategies including the introduction of the vaccines.

16.11.2 Nationwide Acute Febrile Illness (AFI) Study

Another community-based surveillance is currently ongoing at 10 sites of the DRIVEN network (Table 16.2) to estimate the incidence and sero-prevalence of acute febrile illness with a focus on Dengue and Chikungunya viruses (four added infectious agents—typhoid, leptospira, malaria, and scrub typhus)—a prospective multi-centric nationwide cohort study.

Common study protocol, harmonized documents like SOPs, Informed Consent Documents (ICDs), Participant Information Sheets (PIS), Case Record Forms (CRFs), blood sample management and testing manual, and training manuals have been developed and implemented. The protocol has planned 7562 participants to be enrolled and followed up for a period of 12 months.

16.12 Conclusions

Visionary Missions of the Department like NBM facilitated the development of the necessary "tracks" and "trains" for bringing a new product from pilot to market but also played a crucial role in establishing the foundational infrastructure for enhancing the biotechnology ecosystem [7].

16.13 Health Surveillance Sites: Navigating Crisis and Beyond

Establishing and strengthening such community-based surveillance sites with e-data management platforms have significantly benefitted the industry in conducting late-stage clinical trials. These studies provide a rich source of real-world data for research and study to academicians. Government agencies can leverage these community structures for better surveillance, disease dynamics, disease prevention, and disease control. These aid in effective, inclusive, and accountable responses to public health emergencies.

The initiatives undertaken by the National Biopharma Mission to bolster and build capacity, even prior to the COVID-19 pandemic, were effectively utilized, enabling India to become self-reliant in combating the pandemic. The nation's proficiency in conducting top-notch clinical trials and studies promptly and efficiently was a reflection of its strong scientific and research capabilities. We stand prepared to generate real-world evidence in line with international standards. This accomplishment has underscored India's standing as a global frontrunner in vaccine production, demonstrating its potential as a self-sustaining economy that can fulfill its own requirements while making significant contributions to global health security.

Competing Interest The authors have no conflicts of interest to declare that are relevant to the content of this chapter.

References

1. National Biopharma Mission. Catalyzing Transformative Impact On Affordable Product Development in India. Impact Report 2024.
2. Clinical Research and Validation Management Framework (CRVMF). https://www.birac.nic.in/nbm/uploads/2019/08/crvmf.pdf
3. Environment, Occupational Health and Safety Management Framework (EMF). https://www.birac.nic.in/nbm/uploads/2019/08/emf.pdf
4. Nair D, et al. Sero-surveillance to monitor the trend of SARS-CoV-2 infection transmission in India: study protocol for a multi-site, community based longitudinal cohort study. Front Public Health. 2022;10:810353.
5. Sharma N, et al. Community dynamics and engagement strategies in establishing demographic development and environmental surveillance systems: a multi-site report from India. Healthcare (Basel). 2023;11(3):411.
6. Sharma N, Lahariya C, editors. Monograph on setting up demographic, development and environmental surveillance sites. Handbook for Epidemiologists and Public Health Practitioners; 2023.
7. Galal R, Garcia AF. June 06, 2023. Building a home grown biopharma sector in developing countries: The case of India. https://blogs.worldbank.org/en/endpovertyinsouthasia/building-home-grown-biopharma-sector-developing-countries-case-india

Fostering Industry–Academia Partnership

17

Kavita Singh

17.1 Background and Introduction

As the COVID-19 pandemic unfolded, India's strategic response was marked by an innovative collaboration between academia and industry, backed by proactive governmental initiatives.

Joint efforts between industry, academia, and government have been pivotal in advancing scientific research and accelerating the development of crucial healthcare products during the pandemic. The pandemic sparked a remarkable surge of cooperation between Indian academia and industry, supported by the government's enhanced emphasis on research and development to effectively manage the outbreak. These collaborations were instrumental in bridging gaps, leveraging the unique capabilities of each sector, and mitigating weaknesses to swiftly achieve common goals. With a focus on complementing domain expertise, balancing resource strengths, and supplementing existing capacities, these partnerships varied in scope and formality, engaging different mixes of groups to drive innovation, scale-up, regulatory approvals, and deployment. The pandemic-induced surge in cooperation between Indian academia and industry, supported by the government's intensified focus on research and development, was key in developing vaccines, treatments, and diagnostics swiftly, showcasing a united effort to manage the crisis effectively and left behind varied lessons.

Many of these collaborations played a crucial role in quickly developing vaccines and diagnostic tools. Whenever we attempt toward academia–industry partnerships, it is essential to appreciate the unique missions and objectives of these sectors to maximize the benefits of such collaborations. The industry aims to create products that meet societal needs and generate revenue, whereas academia is more

K. Singh (✉)
Drugs for Neglected Diseases Initiative, New Delhi, India
e-mail: ksingh@dndi.org

© The Editor(s) (if applicable) and The Author(s), under exclusive license to Springer Nature Singapore Pte Ltd. 2025

R. Swarup et al. (eds.), *At War with the Single Strand*,
https://doi.org/10.1007/978-981-96-8122-8_17

adaptable, focuses on pioneering new technologies and educating students but often lacks the capability and understanding to bring these innovations to market. Government bodies enhance public health and innovation by funding research and regulating the use of medical technologies. And through their unique strengths, these partnerships provided the much required healthcare tools for managing the pandemic.

This chapter delves into the specifics of these collaborations, highlighting the unique examples of partnership and innovation. The chapter also explains how the learnings of the ongoing Industry–Academia Collaborative Mission for Accelerating Discovery Research To Early Development for Biopharmaceuticals—"Innovate in India (I3) Empowering Biotech Entrepreneurs & Accelerating Inclusive Innovation" allowed the COVID response to pivot on its established systems.

17.2 Government-Led Initiatives Prior to COVID 19 Pandemic

Through the foresight and vision of the government, two missions were launched in 2017 and 2018 by the Department of Biotechnology—the National Biopharma Mission (NBM), entitled "Innovate in India (i3)—Empowering Biotech Entrepreneurs & Accelerating Inclusive Innovation" approved by the Cabinet Committee on Economic Affairs in May 2017 and the Ind-CEPI Mission. These two missions had already made efforts to establish a foundation to build the Biopharma Indigenous Innovation before the pandemic hit us in 2020. These included that while innovations were being supported, the ecosystem was also being facilitated nudging the country's readiness for Vaccine Research and Development. Before the COVID pandemic hit, these two ongoing missions were already supporting the development of innovative vaccine candidates for Influenza, Pneumococcus, Dengue, and Chikungunya. The mission had also mapped the absent facilities and common capacities required for product development which included many facilities to accelerate the biological product development and had identified the unique mechanism of establishing consortiums between academic institutes and industry so as to expedite the discovery and development of novel products in the near future.

These shared facilities at academic institutes were developed to provide access to high-throughput analytical assays, animal models and specialized tests for characterization, production and release, immunogenicity assessment, and clinical assay development to start-ups, with MSMEs at nominal costing. These facilities were also accessible to large pharma and biotech industries, some of the examples being the Imagineering Lab at IIT Kanpur which supports the innovators for metal prototyping and pilot batch production at Yenepoya University, a tissue-based and metal implant prototyping facility working with industry partners and microfluidic device prototyping and pilot batch production facility at the centre for cellular and molecular platform. Similarly, to support the development of biologicals, a microbial cell line repository was established at the Institute of Microbial Technology, Chandigarh, analytical characterization lab at Indian Institute of Chemical Technology, IICT Hyderabad, and at Entrepreneurship Development Centre, Pune. And to support the immunogenicity studies of vaccines, a National Immunogenicity and Biologics Evaluation Centre at IRSHA, Bhartiya Vidyapeeth and a Bacterial Clinical Immunogenicity Lab at Kempe Gowda Institute of Medical Sciences (KIMS), Bangalore were established. The five translational research consortia for Dengue, Chikungunya, Hepatitis E (02), and malaria provided access to the characterized serum samples, viral repository of circulating viral strains, animal models for drug and vaccine trials, and high-throughput assays.

17.3 Government-Led Initiatives for COVID-19 Pandemic

As soon as WHO declared COVID 19 as a Public Health Emergency of International Concern (PHEIC), the Department of Biotechnology (DBT), under the Ministry of Science and Technology, could quickly mobilize and scale-up the resources because of the ongoing missions. At the outset of the pandemic, the Indian government, through various ministries and agencies such as the Department of Biotechnology (DBT) under the Ministry of Science and Technology, Indian Council of Medical Research under the Ministry of Health, and Council of Scientific and Industrial Research under the Ministry of Science and Technology spearheaded initiatives to form a cohesive response framework. These efforts were aimed at bringing together the best of both academia and industry to foster innovations crucial for public health emergencies.

The strategy emphasized critical areas such as diagnostics, therapeutics, vaccines, and other tools, forming the backbone of India's scientific attack on the virus.

17.4 Academia Industry Collaboration for Vaccine Development

Globally, at least nine different technology platforms and about 183 COVID vaccines were under clinical development to create an effective vaccine against COVID-19. With no prior vaccine in market and no prior success story, Indian government science agencies also decided to support many technologies as a risk mitigation strategy. At that time, about 30 different academic organizations and vaccine companies in India were working on nearly eight different platforms with an aim to make available vaccines against SARS-CoV-2. The various platforms supported by the government were:

- Nucleic Acid: DNA and mRNA vaccines
- Protein subunit vaccines
- Inactivated vaccines
- Viral vector
- Vesiculovirus vector

It was imperative that the COVID vaccine development and manufacturing is taken up in a Mission mode, and to ensure a steady supply of vaccines, DBT initiated *Mission COVID Suraksha* with an end-to-end focus from preclinical development through clinical trials, manufacturing, and scale-up development of the prioritized vaccine candidates that would consolidate all available and funded resources toward a warpath for accelerated product development.

There was an understanding that for advancing vaccine development, it would be imperative to support the ecosystem. Considering the number of vaccine candidates under development, it was critical that harmonized processes are followed for appropriate comparisons of immune response elicited by each candidate.

To enter clinical development, COVID-19 vaccine candidates had to be tested in animals, demonstrating safety and efficacy, particularly against a challenge with SARS-CoV-2. Several animal models of SARS-CoV2 were developed at research institutes, including mice expressing ACE2, mouse models with mouse-adapted SARS-COV2 strains, ferrets, and hamsters. The hamster model could mimic severe disease, as is seen in a proportion of infected humans. With multiple vaccine candidates based on varied platforms, streamlining access to animals was critical as it decreased time delays in generating high-quality data.

In addition to animals, the major sources of variability are the assays for estimating immune responses. The assays measuring immune response are complex, require BSL-3 facility, and involve testing of both antibody response and functional response using wild and pseudovirus and cell-mediated immunity (CMI). The research institutes understood the critical need to establish and standardize these assays to measure the immunogenicity of various COVID 19 vaccine candidates for supporting vaccine companies in submitting data to regulatory agencies.

These National Service Facilities for conduct of animal studies and immunological assays were available to COVID-19 vaccine developers that were established by academic institutes. Existence of such common facilities enables standardized access to animals and assays, thereby accelerating the development of COVID-19 vaccines.

The two examples being highlighted are as follows:

1. The facility established at IRSHA under National Biopharma Mission developed (1) standard crystal violet staining-based virus neutralization test against wild-type, Delta, Kappa, Omicron BA.1, BA.5, and XBB strains, (2) ACE2 receptor-based neutralization assay for SARS-CoV-2 variants in 25 uL sample volume using MesoScale Discovery (MSD) platform, and (3) luciferase reporter-based pseudovirus SARS-CoV-2 neutralization test against wild-type strains for immunogenicity evaluation of SARS-CoV-2 vaccines in clinical trial. They provided

services to Indian Vaccine manufacturers like Zydus Cadila, Gennova Biopharmaceuticals, Bharat Biotech, Biological E, CMC Vellore, and Reliance Life Sciences and international clients also. More than 18,000 samples were tested at NIBEC, IRSHA for PRNT assay.

2. The Bioassay Laboratory (BL) of THSTI is the only NABL-accredited (ISO 17025:2017) cGLP laboratory in the Department of Biotechnology Institute. Under the Coalition for Epidemic Preparedness Innovation (CEPI) network, BL provided testing services for Phase I–III samples to vaccine developers such as Biological E Ltd.; S K Bioscience, South Korea; The Government Pharmaceutical Organization (GPO), Thailand; Aurum Institute, South Africa; Medigen Biotechnology Corp., Taiwan; Nanogen Pharma, Vietnam; and Litevax, The Netherlands and also provided collaborative services for WHO Solidarity Trial, Geneva, Switzerland. Even now, BL is actively involved in establishing the standards and controls for CEPI assays for the CEPI network laboratories. Bioassay lab also collaborated with the World Health Organization (WHO) and National Institute for Biological Standards and Control (NIBSC) to establish the second international standard for anti-SARS-CoV-2 immunoglobulin, a reference panel for SARS-CoV-2 variants of concern and to establish the first WHO International Standard for SARS-CoV-2 antigen. BL has also developed the platform for binding antibody assay and antiviral screening and neutralization assay platform for flaviviruses such as Dengue, Zika virus, Japanese encephalitis virus, and West Nile virus.

17.5 Academia Industry Collaboration for Diagnostic Development

Indian academia and companies were involved in developing several in vitro diagnostic kits (IVDs) that were effective in mass testing and identification of infected individuals and epidemiology studies.

Many academia industries worked together to develop a healthy pipeline of diagnostics tools and the reagents required to develop these tools. The kit components like material transfer media, swabs, enzymes for RTPCR, and viral transport media were quickly developed and scaled up.

Some of the academia–industry collaboration examples are mentioned in Table 17.1.

Table 17.1 Some key academia–industry collaborations

S No	Type of organization	Title
1	A commercial company	Cloning and large-scale continuous manufacturing of SARS-CoV2 spike protein for recombinant protein-based diagnostic kit development and to be provided to academia or other diagnostic developer
2	Denovo Biolabs Pvt., Ltd. (collaborator: ICAR-National Institute of High Security Animal Diseases (NIHSAD), Bhopal	Development and validation of LFA POCT kits for SARS-COV-2 antigen detection
3	IIT-Guwahati, collaborators: IIT Jammu, NICED and Stethox Healthcare	CRISPER-based diagnosis of COVID-19 using paper microfluidics
4	IIT Delhi Collaborator: Vanguard Diagnostics Pvt., Ltd.	Development and evaluation of an aptamer-based lateral flow assay kit for detection of SARS-CoV2 detection.
5	SciGenom Labs Private Limited Collaborators: (1) Anamol Laboratories, Navi Mumbai, (2) Christian Medical College, CMC, Vellore	Development of a rapid cost-effective point-of-care SARS-CoV2 COVID-19 detection test
6	SCMS Institute of Bioscience and Biotechnology Research and Development Collaborator—Malabar Cancer Center	LAMP-based, point-of-care (POC) rapid and low-cost diagnostic test for COVID-19
7	Indian institute of technology, Ropar Collaborators: Dayanand Medical College Ludhiana and Parmar Hospital, Ropar	Development of a low-cost attachment, a simple solution that can leverage the AMBU bag to rapidly convert it to an automatic mechanical ventilator with all basic functionalities desired to manage a pneumonia or ARDS patient
8	PGIMER, Chandigarh collaborators: UIET; Panjab University Chandigarh; IIT, SCL, Mohali Punjab	Design, develop, and laboratory test a Smart Mechatronic Artificial Respiratory Therapy System (SMARTS), relevant to clinical management of COVID-19

17.6 Conclusion

The importance of enduring strategic partnerships between research institutes and industrial entities cannot be overstated. Such relationships prioritize mutual benefits by aligning strategic goals, fostering deep technical and strategic connections, and minimizing interaction costs. The pandemic has notably influenced how research institutes and industry partners collaborate. Despite stringent restrictions, these partnerships adapted quickly during the pandemic, initiating urgent projects like vaccine development and diagnostic production. This adaptability was facilitated by trust and a shared urgency, allowing for swift agreement on objectives and preliminary terms, bypassing typical bureaucratic delays. In post-pandemic economic recovery, research institutes with translational research capacities and mechanism to pursue industry academia collaborations are poised to play a crucial role in driving, shifting strategic priorities toward more challenge-led programs that integrate research with commercial application.

Looking forward, the challenge is to leverage the proven capabilities of academia–industry collaborations in addressing immediate crises to foster a sustained innovation-led recovery. This will require maintaining core research funding, enhancing support for translating research to applications, and ensuring sustained investment in partnerships and knowledge exchange frameworks. Such strategic focus will be essential for harnessing the full potential of research institutes in the national recovery efforts post-COVID.

Competing Interest The authors have no conflicts of interest to declare that are relevant to the content of this chapter.

The Start-Ups' Response: Fostering Resilience: The IKP Story of Leveraging Start-Up Ingenuity in Tackling the COVID-19 Pandemic

18

Deepanwita Chattopadhyay, Viswanadham Duppatla, Dedeepya Chermala, Bindu Nishal, and Srikrishna Sulgodu Ramachandra

18.1 The Unprecedented Impact of COVID-19

COVID-19 has cast its shadow over nearly every corner of the planet, with an overwhelming 768 million confirmed cases and the loss of around 6.9 million lives between December 2019 and July 2023 [1]. Yet, these figures only offer a glimpse of the scale and depth of the calamity. With nationwide lockdown, large-scale job losses, and broken supply chain and logistics, the impact of the pandemic extended far beyond immediate health concerns, as it adversely affected global health goals, including maternal and child mortality, HIV and tuberculosis control programs, and overall, the Sustainable Development Goals (SDGs).

As nations united to confront the pandemic challenges, it also served as a reminder of the shared vulnerability that transcends borders and the need for collective responsibility that could shape a resilient future for all. It also served as an opportunity for science-driven technology innovations to help mitigate the COVID-19 pandemic and prepare for similar future occurrences.

D. Chattopadhyay (✉)
IKP Knowledge Park, Hyderabad, Telangana, India
e-mail: deepanwita@ikpknowledgepark.com

V. Duppatla
Biopharma Innovation & Biomanufacturing, IKP Knowledge Park, Hyderabad, Telangana, India

D. Chermala
Biopharma and HealthTech Innovations, IKP Knowledge Park, Hyderabad, Telangana, India

B. Nishal
Agri Innovations and One Health, IKP Knowledge Park, Hyderabad, Telangana, India

S. S. Ramachandra
Healthcare Projects, IKP Knowledge Park, Hyderabad, Telangana, India

© The Editor(s) (if applicable) and The Author(s), under exclusive license to Springer Nature Singapore Pte Ltd. 2025

R. Swarup et al. (eds.), *At War with the Single Strand*,
https://doi.org/10.1007/978-981-96-8122-8_18

18.1.1 Government's Engagement with the Start-Up Ecosystem

The Government of India played a pivotal role in combating the COVID-19 pandemic through a multipronged approach, including the development of COVID-19 vaccines like Covishield (manufactured by the Serum Institute of India) [2] and Covaxin [3] (India's first indigenous COVID vaccine developed by Bharat Biotech) and their distribution across the country through the COWIN portal [4].

The way in which the entire innovation ecosystem of the country was leveraged was unique. Various government agencies engaged with academia, start-ups, MSMEs and corporates, incubators, banks, angels, and VCs through several schemes and public–private partnership models. It stood by the start-up and MSME ecosystems by extending grants, loans, tax relief, and regulatory adaptations, thereby cushioning the impact on emerging enterprises.

Healthcare start-ups in India found crucial support from various government agencies and key stakeholders. The Department of Biotechnology (DBT) [5] and the Biotechnology Industry Research Assistance Council (BIRAC) [6] spearheaded the COVID Suraksha program and introduced a range of schemes and grants aimed at propelling advancements in diagnostics, vaccines, therapeutics, and allied technologies. Concurrently, the Indian Council of Medical Research (ICMR) [7] and the Central Drugs Standard Control Organisation (CDSCO) [8] streamlined the approval processes for COVID-19 drugs, vaccines, and medical devices, while the Ministry of Health and Family Welfare (MoHFW) [9] issued telemedicine practice guidelines, empowering registered medical practitioners to extend remote consultations via technology platforms. This in turn opened up new avenues for start-ups to build solutions and platforms. The Department of Science and Technology (DST) [10] launched the Centre for Augmenting WAR with the COVID-19 Health Crisis (CAWACH) [11] program and funded start-ups with innovative pandemic-centric technologies. Complementing this, the Ministry of Electronics and Information Technology (MeitY) [12] facilitated innovation by initiating an innovation challenge, fostering the creation of secure and cost-effective videoconferencing platforms. The Start-up India's COVID-19 Resource Section offered tailored support for start-ups, investors, and mentors, along with the Government e-Marketplace dedicated COVID-19 platform streamlined supplier listings. The Atmanirbhar Bharat Abhiyan unveiled a series of substantial relief packages, amounting to approximately 8% of GDP, encompassing fiscal stimulus, tax mitigation, credit guarantees, liquidity enhancements, and regulatory adjustments.

These collaborative and supportive measures, and all these in an unprecedented fast-track mode, collectively fortified innovation, expedited vital solution development, and empowered healthcare start-ups to effectively navigate pandemic-induced challenges.

18.2 The Role Played by IKP Knowledge Park

IKP Knowledge Park (IKP) [13] is a 200-acre Science Park and Incubator with a physical presence in Hyderabad and Bangalore, India. IKP promotes the advancement of technology-based innovators, entrepreneurs, and small and large companies through customized space, shared equipment, incubation, mentorship, advisory services, and funding. IKP has so far supported over 1500 companies from seven countries, 90% of which are start-ups.

IKP's campus at Genome Valley, Hyderabad, which houses several large pharma and biopharma R&D centers and contract research organizations started adopting stringent safety and preventive measures in the campus much before the COVID-19 lockdown was announced in India. IKP and its facilities came under emergency services, and the management ensured that it was open 24/7 throughout the lockdown period after the first day of mandatory closedown. This was not only critical for the companies at IKP but also served in housing an MSME to produce sanitizing solutions which it was unable to do otherwise.

The challenge for the country was that health infrastructure was not prepared to deal with the COVID-19 situation. Pharmacies ran out of sanitizing solutions, masks, and disinfectants. Primary healthcare workers did not have access to Personal Protective Equipment (PPE). Hospitals did not have sufficient numbers of ventilators in place or ICU beds. Screening and diagnostics protocols had to be set up and supply ensured. The challenge for IKP was on the one hand ensuring the survival of the IKP start-ups in such extraordinary circumstances and on the other hand to devise a mechanism to help reduce the acute shortfall of preventive, screening, diagnostics, and disease management solutions.

Being part of the larger innovation ecosystem, IKP Knowledge Park launched the IKP COVID Fund (I-CO Fund) [14] on March 30, 2020, in less than a month after WHO declared COVID-19 a pandemic. The I-CO Fund was created with the aim to empower start-ups and MSMEs with essential working capital and enable the selected entrepreneurs to fine-tune their near-ready solutions for manufacturing and connect with manufacturing facilities and supply-chain networks to help in the production, deployment, and distribution of their innovations. The starting amount for the fund was INR 2.0 Cr from IKP Trust, and it also received support from Tynor Orthotics Pvt., Ltd., Hindustan Syringes Pvt., Ltd., and a few individual donors. While much larger funds were put together for COVID-relevant technologies as the scale and impact of the pandemic hit India, the I-CO Fund was unique as it was one of the first responders to set up a catalytic fund and believed in the innovation ability of our start-ups and SMEs to respond to such a calamity.

The first Call for Proposals was announced on March 30, 2020, and selection was completed for the first cohort, and funds were disbursed by the third week of April 2020. This was followed by a second Call in May 2020. The idea was to be agile and respond quickly to the crisis. While the amount was modest, it served as a catalytic fund and helped start-ups to fine-tune their products/technologies toward addressing the COVID-19 crisis. On seeing the positive response, BIRAC partnered with the I-CO Fund and offered a matching support of up to INR 2.5 Cr. A subsequent Call was announced in May 2021.

The I-CO Fund selected 20 start-ups and MSMEs to refine and repurpose 19 technology solutions for manufacturing and facilitate their access to production facilities and supply chain networks (Table 18.1). This support enabled the creation, deployment, and distribution of 17 innovative solutions from 18 enterprises across various domains that are listed in Table 18.1 [15]. Two start-ups could not meet the development milestones.

Table 18.1 List of start-ups supported through the I-Co Fund

Category	S. No.	Name of the company	Innovation
Personal	1	Aarna Biomedical Products Pvt., Ltd.	Suraksha PPE kit
	2	Butterfly Edufields Pvt., Ltd.	Face shields and protective intubation hoods
Protective equipment	3	Emflux Motors Pvt., Ltd.	Reusable face shields
	4	Parisodhana Technologies Pvt., Ltd.	SHG-95: A hybrid, multiply reusable face mask (billion social masks)
Disinfection	5	MicroGo LLP	GOassure, GOsafe
	6	Strl BioSystems Pvt., Ltd.	Strl Air: Continuous and persistent air disinfection system
Screening and monitoring	7	DocketRun Tech Pvt., Ltd.	DocketRun Traq-19: AI-enabled surveillance system for public places
	8	Evelabs	Dripo
	9	Monitra Healthcare Pvt., Ltd.	upBeat: 24x7 heart monitoring device
	10	Predible Health Pvt., Ltd.	LungIQ: An Artificial Intelligence tool to help doctors assess the degree of COVID-19 severity and impact on the lungs of patients from CT scan images
	11	Tinkerbee Innovations Pvt., Ltd.	TempBee: Battery-assisted/operated temperature monitoring device
	12	Turtle Shell Technologies Pvt., Ltd.	Dozee: Noncontact respiratory monitoring device
Diagnostics	13	AffigeniX Pvt., Ltd.	Covid-19 Q RT PCR testing for molecular in vitro diagnostic test
	14	Dhwani LifeSciences Pvt., Ltd.	Cytokine cocktail derived by nonspecific activation of T-cells for treating COVID-19 patients
	15	Huwel Lifesciences Pvt., Ltd.	COVID-19 RT QPCR diagnostic kit

(continued)

Table 18.1 (continued)

Category	S. No.	Name of the company	Innovation
Medical devices and equipment	16	Janitri Innovations Pvt., Ltd.	KEYAR: A portable labor monitoring device
	17	Jeevtronics Pvt., Ltd.	SanMitra 1000 HC: Hand-cranked defibrillator
	18	Remidio Innovative Solutions and Biodesign Innovation Labs	RespirAID ventilator
	19		
Online counselling services	20	XEqualsZERO Pvt., Ltd.	TickTalkTo: Telephone and online-based counselling services to battle psychological crisis due to COVID-19

18.3 Notable COVID-19 Solutions and Start-Ups Backed by IKP

18.3.1 Suraksha PPE Kit (Aarna Biomedical Products Pvt., Ltd.)

Established in 2017 by Dr. Pawan Mehrotra, a distinguished cancer biologist, Aarna Biomedical Products Pvt., Ltd. emerged from a vision conceived in 2014. Driven by the aspiration to offer an affordable and accessible solution for breast cancer patients in India compelled to undergo mastectomies, the venture conceived the "Poorti-Kit". This post-mastectomy kit has been meticulously developed to cater to the needs of breast cancer patients, encapsulating the essence of the firm's commitment to enhancing the lives of patients in a pragmatic and compassionate manner. Introduced in April 2020 with the collaborative support of IKP, BIRAC, and IIT-Kanpur, the SURAKSHA PPE Kit stands as a comprehensive solution designed for safeguarding personnel operating in vulnerable environments (Fig. 18.1). This disposable, full-body wearable encompasses vital components such as a face shield, disposable bag, shoe covers, a head-covered body coverall, an outer poly pack for UV sanitization, and nitrile hand gloves for medical examination. Available in distinct versions for both health and non-health workers, the kit has achieved a significant milestone, with over 15,000 units sold. In January 2021, this innovative kit garnered recognition by being selected for a direct presentation to the Prime Minister, Shri Narendra Modi, during the International Prarambh Summit.

18.3.2 SHG-95 Masks (Parisodhana Technologies)

Parisodhana Technologies Pvt., Ltd. from Hyderabad has developed SHG-95® (Billion Social Masks), a hybrid multi-ply face mask with high particulate filtration as per the ASTM F2299 standard (PFE > 90%) and bacterial filtration efficiency as

Fig. 18.1 Components of Suraksha kit

per the ASTM F2101 standard (BFE > 99%). These masks have been designed for optimal comfort and breathability, even in tropical conditions, using handwoven cotton materials with an added specialized filtration layer (Fig. 18.2). These reusable masks come at an affordable estimated cost of INR 50 per piece, catering to the general public's needs. With a substantial sales record of over 145,000 units and an almost equal or higher number of donations/free supply to the frontline warriors during the Covid-19 pandemic, and further supported by funding from Grand Challenges Canada, this initiative not only addressed the challenges of the COVID-19 era but also uplifted the livelihoods of Self-Help Groups (SHGs).

18.3.3 Goassure (MicroGo)

MicroGO, a start-up from Chennai, working on smart solutions in water, sanitation, and hygiene (WaSH) challenges and infection control developed multiple sterilization and disinfectant products like GOsafe, GOassure, GOpure, and GoFresh that were deployed during the COVID-19 period.

MicroGO embarked on its innovation journey in 2016 and received the support of the Biotech Ignition Grant support (BIG) from BIRAC in 2018 for developing a portable, energy-efficient alternative to autoclaves for sterilizing surgical instruments. In 2019, a patent was granted to MicroGO for its Tubelet Technology. This technology is used in its disinfectants and sterilization products. The start-up has

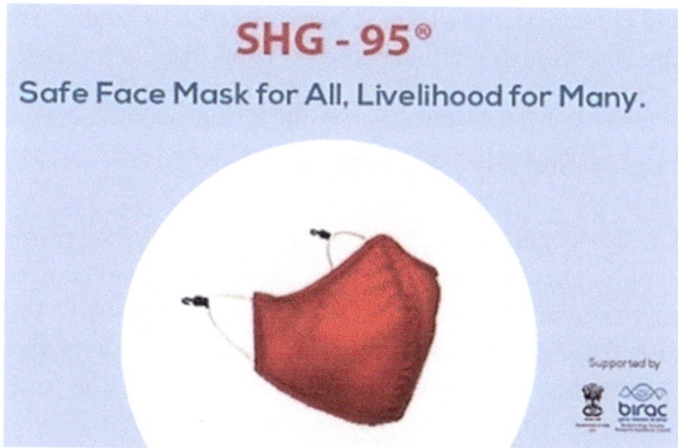

Fig. 18.2 A hybrid multi-ply face mask by Parishodhana Technologies

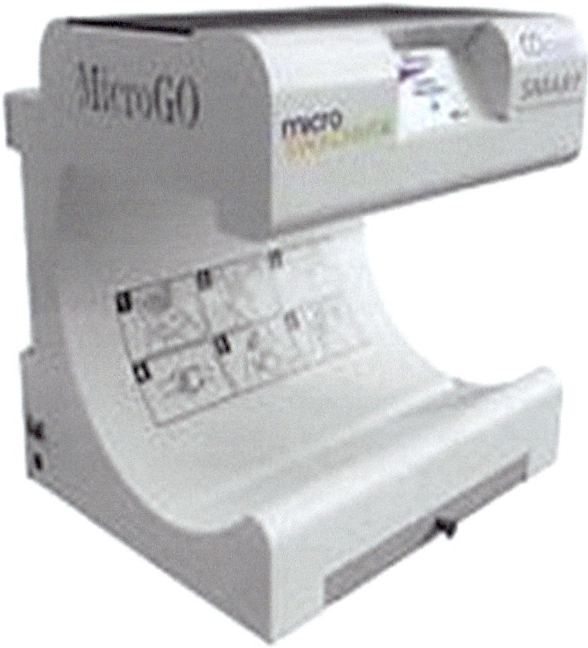

Fig. 18.3 GOassureTM MAX, an ingenious solution that automates hand hygiene

been a recipient of the "Outstanding Woman Entrepreneur in disinfectant & sterilisers" award at India International Science Festival 2020.

GOassureTM MAX is an ingenious solution that automates hand hygiene and assures compliance monitoring to suit clients' requirements (Fig. 18.3). The device ensures 20 s of contactless hand hygiene with an optimal volume of alcohol-based hand rub. The IoT-enabled device has an in-built LCD display and RGB lighting indicators for user guidance and ease of use.

MicroGO has installed hundreds of GOassure units at 19 airports under Airports Authorities of India including Chennai, Kolkata, Pune, Baroda, Hyderabad, Jaipur, and Guwahati, in hospitals, corporate offices, and in a remote area in Rajasthan. The device was also deployed at the Prime Minister's Office.

18.3.4 Strl Air (Strl BioSystems)

Strl BioSystems Pvt., Ltd., a dynamic social enterprise, is effecting positive change in public health and well-being through the provision of secure and hygienic indoor environments (Fig. 18.4). Going beyond innovative product creation for mitigating infectious disease transmission, the company is actively equipping individuals to elevate their quality of life. Guided by the vision to democratize access to sterile, germ-free spaces, they harness the power of biotechnology and nanotechnology. Their endeavours encompass pioneering disinfection and infection control systems, engineered to incessantly neutralize pathogens like the novel Coronavirus, both in airborne and surface settings, marking a significant stride toward enhanced public health.

Their offerings tailored for COVID-19, Strl BioSystems Pvt. Ltd., include:

1. *Strl AIR™:* An innovative bio-sterilizer featuring a proprietary catalyst that ionizes the air, producing reactive oxygen species capable of exterminating microbes. This compact, user-friendly, and portable solution has undergone rigorous testing and validation by esteemed Indian institutions.
2. *AirSPD:* A versatile device, AirSPD, employs plasma to sterilize surgical instruments and personal protective equipment. Notable for its simplicity, speed, effectiveness, and eco-friendliness, this solution ensures thorough sterilization.

Meticulously engineered to combat even the most resilient pathogens, these products prioritize individual safety and environmental cleanliness. Additionally, they offer affordability and demand minimal maintenance, exemplifying a harmonious blend of efficacy and user convenience.

Fig. 18.4 Strl BioSystems provides solutions for a germ-free environment

18.3.5 Traq-19 (DocketRun Tech Pvt., Ltd.)

DocketRun Tech Pvt., Ltd. is a start-up specializing in artificial intelligence solutions within computer vision. Established in 2018 by Dr. Ajay Kabadi, Chetan Kulkarni, and Shweta Shettar, the company is headquartered in Hubballi, Karnataka.

Responding to the challenges of the COVID-19 pandemic, DocketRun Tech Pvt., Ltd. offers valuable services tailored to these circumstances. Their innovative COVID-19 solution—Traq-19—is a robust video analytics platform that helps businesses monitor and manage pandemic-related safety protocols, including social distancing, mask compliance, occupancy limits, and temperature monitoring.

In addition, the company provides sophisticated "face recognition" and "object detection" systems for accurate identification and verification from live video feeds or images.

18.3.6 upBeat (Monitra Healthcare)

Monitra Healthcare Pvt., Ltd. based in Hyderabad is a pioneering start-up founded in 2016. Specializing in biosensing and medical devices, the company's flagship product, upBeat®, is a wire-free ECG monitor that aids clinicians in diagnosing complex heart conditions like unexplained fainting and palpitations (Fig. 18.5). This remote 24x7 heart monitoring device offers early detection of cardiac ailments, including arrhythmias and atrial fibrillation, and has garnered support from BIRAC, IKP, USISTEF, and Indian Angel Network.

Amidst the COVID-19 pandemic, upBeat® facilitated cardiac monitoring for individuals afflicted with heart conditions, emerging as a vital remote heart monitoring solution tailored for segments like senior citizens undergoing quarantine and social isolation. Throughout this challenging period, upBeat® effectively empowered healthcare practitioners by enabling continuous heart rhythm tracking for their patients, even under the constraints of stringent social distancing measures.

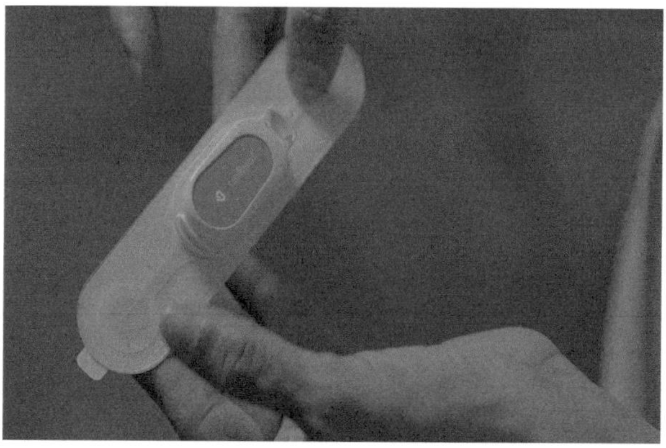

Fig. 18.5 upBeat®, a wire-free ECG monitor helping doctors diagnose heart conditions

18.3.7 LungIQ (Predible Health Pvt., Ltd.)

Predible Health Pvt., Ltd., headquartered in Bengaluru, a pioneering start-up harnessing the power of artificial intelligence (AI) in radiology and specializing in lung- and head-related ailments, responded to the challenges posed by the COVID-19 pandemic, through LungIQ, a cutting-edge software adept at assisting healthcare practitioners in the pre-screening and identification of COVID-19 patients through the analysis of thoracic CT scans, pinpointing lung damage indicators (Fig. 18.6). LungIQ reports ensure high-quality reporting by using quantitative indicators, visual representations, and the most recent clinical standards. It is suitable for all CT scanner types. The tool can also assist in identifying severe cases and eliminate subjectivity in reporting.

This technology has been deployed across 15 sites, contributing to the diagnosis of various lung conditions. Predible Health was acquired by Nference in 2022.

18.3.8 TempBee (Tinkerbee Innovations Pvt., Ltd.)

Tinkerbee Innovations from Bengaluru specializes in the development of intelligent sensors, actuators, energy-efficient microcomputing devices, and advanced wireless networks. Their flagship product, known as TempBee, serves as a compact device tailored for cold storage applications (Fig. 18.7). Designed explicitly for freezer-type environments, it provides real-time updates on temperature fluctuations within its equipped zone. With a temperature measuring range spanning from −50 °C to 125 °C, the device ensures comprehensive coverage. It boasts auditory and visual alert systems that promptly notify users when temperatures surpass or fall below the predetermined threshold, facilitating immediate corrective measures. TempBee supports both Wi-Fi and LoRaWAN connectivity. Its battery-powered design further enhances convenience, and installation is straightforward. This device is offered in distinct variations tailored to suit diverse needs, encompassing server rooms, deep freezers, refrigerators, and transit containers.

During the 2021–2022 period, TempBee received support through the I-CO Fund's Call 3. This played a pivotal role in deploying the device for monitoring

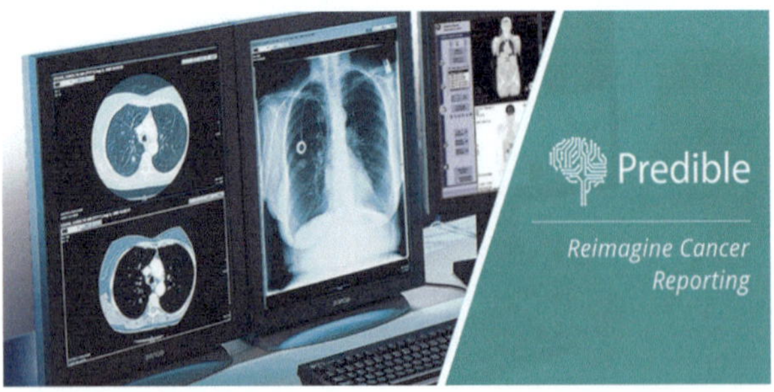

Fig. 18.6 LungIQ is a cutting-edge software that assists healthcare practitioners for pre-screening and identification of COVID-19 patients

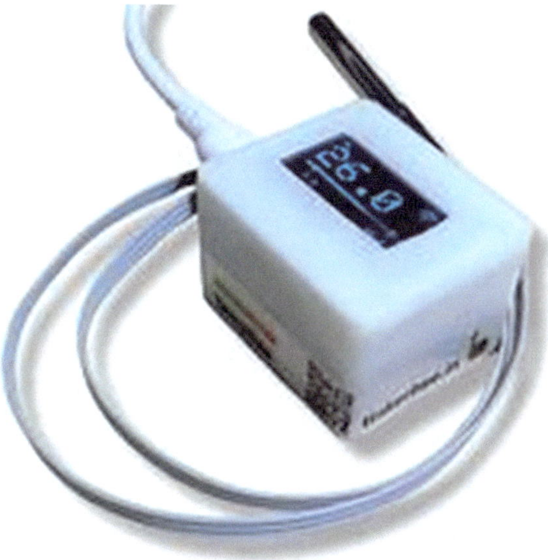

Fig. 18.7 TempBee offers tracking of real-time variation in temperature

temperature-sensitive cold chains used for storing medicines and vaccines. Hospital pharmacies embraced the technology, installing TempBee units to vigilantly oversee temperature conditions within medicine storage units.

18.3.9 *Dozee* (Turtle Shell Technologies)

Turtle Shell Technologies Pvt., Ltd., founded in 2015 and headquartered in Bengaluru, is the visionary force behind Dozee—a transformative device designed to monitor vital signs in a contactless manner (Fig. 18.8). This ingenious solution seamlessly converts any bed into a high-precision health monitor, capturing essential metrics like heart rate, respiration, sleep patterns, and stress recovery with exceptional accuracy. Dozee's utility expanded to include remote monitoring of COVID-19 patients in quarantine centers and hospitals across India. The device's unobtrusive nature is facilitated by its slim sensor sheet positioned beneath the mattress, which comprehensively tracks health indicators including oxygen saturation. These valuable data are effortlessly accessible via an intuitive app, which not only presents health insights and trends but also facilitates data sharing with healthcare professionals and family members. Dozee's ease of use, ability to integrate seamlessly, and its affordability allowed deployment in several government health facilities.

Dozee is ISO 13485:2016- and IEC 60601-1-2d- certified and notified by CDSCO. Dozee received US FDA clearance in December 2022. It has earned several awards including the Most Admired and Preferred HealthTech Emerging Unicorn at the 127th Conference of Unicorns and Emerging Unicorns and 2021 Awards by CE Worldwide.

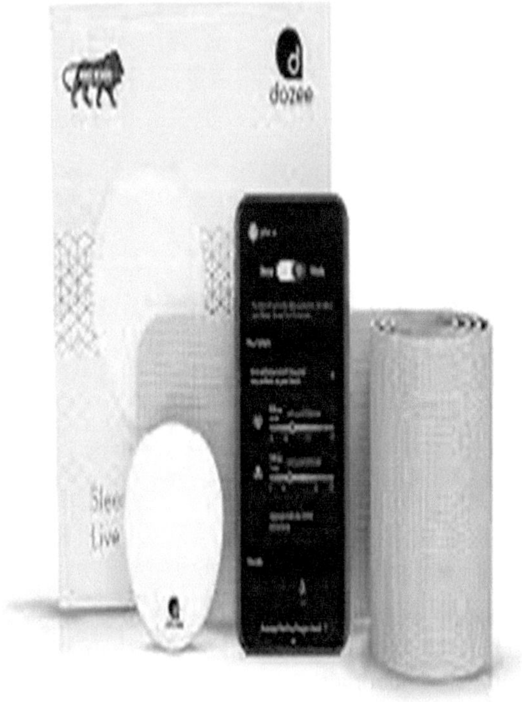

Fig. 18.8 Dozee monitors vital signs in a contactless manner

18.3.10 Molecular Diagnostic Kits (AffigeniX Pvt., Ltd.)

AffigeniX Pvt. Ltd. is a Chennai-based start-up within the CREW group's life science sector, specializing in tailoring biosolutions for Biopharma, Biotech, Food, and Agriculture industries. With a notable focus on therapeutic and diagnostic advancements, the company actively engages in combating metabolic, autoimmune, oncological, and targeted tropical diseases. The portfolio of AffigeniX encompasses an array of offerings, ranging from biochemistry reagents and blood grouping sera to turbidimetry and molecular diagnostics reagents.

Amid the challenges presented by the COVID-19 pandemic, the company showcased its agility and expertise by developing a molecular diagnostic kit designed for the accurate identification of COVID-19. Employing the one-step real-time reverse transcriptase PCR with hydrolysis probe chemistry, the kit targets specific and conserved segments of the SARS-CoV-2 RNA, including the E gene, N1 gene, and ORF1ab gene fragments (Fig. 18.9). The kit has received approval from the Indian Council of Medical Research (ICMR).

Fig. 18.9 AffigeniX offers one-step real-time RT-PCR

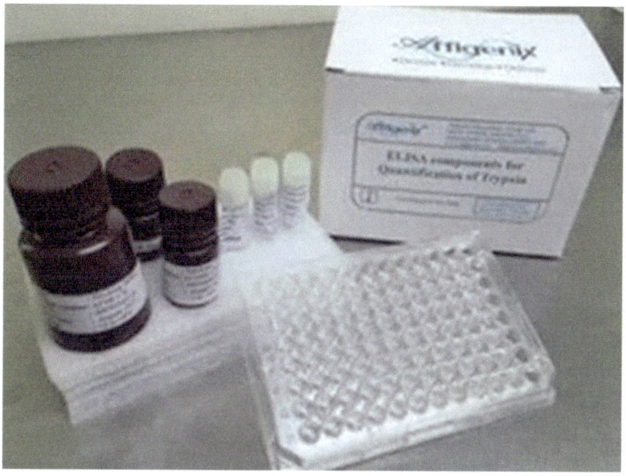

18.3.11 RT-PCR Kits (Huwel Lifesciences)

Huwel Lifesciences Pvt., Ltd., founded in February 2015 in Hyderabad, is dedicated to the development, manufacturing, and distribution of molecular reagents and diagnostic kits. Throughout the COVID-19 pandemic, the company demonstrated its commitment by offering dependable, top-tier diagnostic services to medical professionals. Huwel Lifesciences developed the Quantiplus COV kit for COVID-19 testing (Fig. 18.10). The kit showed high efficiency and could produce results in a short period of 2 h. The kit was made using a single formulation with no requirement to add separate components while setting up reactions. It reduced the time of reaction and also avoided contamination. Approvals were received from ICMR for commercial use of the kit on April 10, 2020.

I-CO Fund helped Huwel scale up its laboratory facility to meet the demand. The facility is outfitted with state-of-the-art clean rooms, purposefully designed to accommodate the development and manufacturing of PCR, real-time PCR kits, and rapid real-time PCR machines, complemented by essential reagents. Adhering to stringent Good Manufacturing Practices (GMPs) and employing meticulous quality control procedures, Huwel maintains a high level of documentation, ensuring the consistent delivery of top-notch kits.

18.3.12 Keyar CM and Keyar DT (Janitri Innovations Pvt., Ltd.)

Janitri Innovations Pvt., Ltd., headquartered in Bengaluru, is a pioneering start-up dedicated to advancing maternal and child healthcare through sustainable solutions. Since its establishment in 2016, the company has been driven by

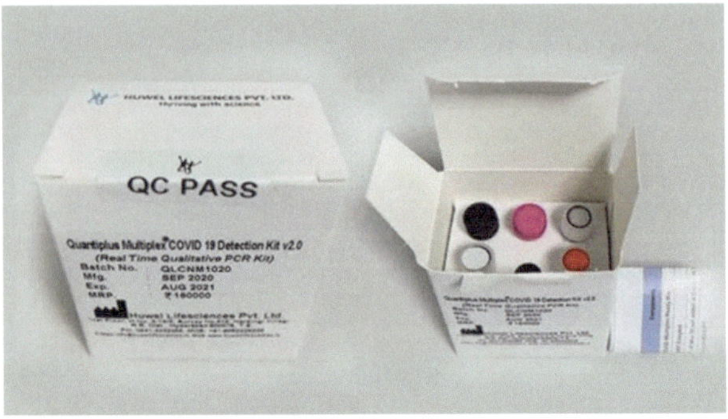

Fig. 18.10 Quantiplus COV kit for COVID-19 testing

Fig. 18.11 Keyar: A portable labor-monitoring device

the mission to ensure maternal and infant well-being. Their comprehensive range of customizable antenatal and labor-monitoring devices include Keyar CM for fetal and maternal health tracking, Keyar DT for wireless fetal monitoring, Keyar DT Lite for home-based high-risk monitoring, Keyar mini for pocket-sized fetal heart rate measurement, Daksh for pregnancy monitoring software, and the Mother's App for smart pregnancy tracking and reporting (Fig. 18.11).

The solution allowed remote monitoring of pregnancy by doctors, especially useful at the time of lockdowns and social distancing. The fetal heart rate and other health reports could be sent to doctors through the app, making the solution highly efficient and useful.

18.3.13 SanMitra 1000 HCT—Hand-Cranked Defibrillator (Jeevtronics Pvt. Ltd.)

Based in Pune, Jeevtronics Pvt. Ltd. is a unique company that specializes in innovative medical devices, particularly defibrillators, designed for resource-limited settings. Defibrillators restore normal heart rhythm by delivering a high-voltage shock to cardiac arrest patients. Their creation, the SanMitra 1000 HCT, is a hand-cranked defibrillator capable of administering a biphasic shock within 15 s of cranking. Operating on both AC mains and manual power, it offers heightened reliability and affordability, particularly in regions with unreliable electricity.

The company has successfully distributed its defibrillators to hospitals and healthcare centers across India, importantly in tier II and III towns. Certified by the Bureau of Indian Standards and the Central Drugs Standard Control Organization, the SanMitra 1000 HCT has been used effectively in Covid-19 wards, especially in Maharashtra and Chattisgarh Covid centers, demonstrating its significance in patient resuscitation efforts. The company's impact extends internationally, with inquiries about exporting its product to other nations.

18.3.14 RespirAID Ventilators (Remidio Innovative Solutions and Biodesign Innovation Labs)

Remidio Innovative Solutions and Biodesign Innovation Labs, based in Bengaluru, India are healthcare start-ups united by a commitment to affordable and inventive medical solutions, particularly tailored for resource-limited settings.

Remidio, founded in 2011, holds ISO13485 certification and employs design thinking to enhance healthcare accessibility. Their flagship product, a portable fundus camera, detects eye conditions like diabetic retinopathy. Biodesign, established in 2017 by Adithya Pasupuleti and Gautham Pasupuleti, envisions democratizing healthcare. Their innovative RespirAID ventilator is cost-effective and adaptable, catering to ER, ambulance, and postoperative care scenarios. Both start-ups have risen to the challenges of the COVID-19 pandemic, supplying vital medical tools and solutions, with partnerships amplifying their impact across India and beyond.

18.3.15 Tick TalkTo (XEqualsZERO Pvt. Ltd.)

TickTalkTo is an mHealth App developed by Mr. Abhay Singhal to leverage technology to bridge the gap between individuals and mental health professionals, enabling online counselling and psychiatric support. He later incorporated XequalsZERO as a start-up in 2019 in Ludhiana, Punjab. Beyond its core functionality, the app extended free counselling assistance to frontline healthcare workers and those grappling with the far-reaching impact of COVID-19, helping them navigate through trauma, loss, and isolation challenges.

In 2020, the venture secured support from the I-CO Fund, facilitating the creation of a dedicated mental health dashboard tailored for India. Additionally, the endorsement from the CAWACH initiative of NSTEDB, DST, Government of India, further highlighted the initiative's significance. Recognized as a COVID-19 support solution by Accelerating the Growth of New India's Innovations (AGNii), a National Technology Commercialization Program, TickTalkTo firmly established itself as a pioneering resource.

IKP also participated in the CAWACH initiative of DST as a CAWACH satellite center for the implementation of the program and supported eight start-up ventures. Notably, MicroGO, Predible Health, and Dozee (Turtle Shell Technologies) teams were able to deploy their solutions as they had already validated their products through the I-CO Fund.

18.4 Conclusions

The rapid deployment of I-CO Fund in the wake of the COVID-19 outbreak demonstrated a timely and impactful response to the prevailing health crisis in India. With the nation grappling with both health and economic challenges, the fund's establishment provided support for start-ups. Beyond financial assistance, IKP's comprehensive approach, encompassing incubation support, technical guidance, mentorship, industry connects, and fundraising help, emerged as a robust foundation for these enterprises to weather the storm. It cultivated a nurturing ecosystem that incubated inventive solutions aimed at tackling the issues arising from the pandemic. Notably, collaborations with diverse stakeholders and initiatives, including BIRAC co-funding and CAWACH, fortified the supportive ecosystem, enhancing the collective response to COVID-19 challenges and in shaping a resilient future.

To better understand the challenges of a post-pandemic world and strengthen the country's One Health vision, IKP launched the International One Health Incubator (IOHI) in October, 2021. The goal of this virtual–global incubator is to identify and provide support to science-driven technology-based solutions to adopt a holistic One Health approach to solving global challenges by keeping in mind the concerns of the health of humans, animals, and the planet. The support to start-ups includes support for fundraising, connecting with the validation partners, national and global deployments, IP and regulatory needs, etc., but most importantly, evaluating and monitoring them so that they remain on the One Health trajectory to solving problems and abide the OH approach.

Competing Interest The authors have no conflicts of interest to declare that are relevant to the content of this chapter.

References

1. https://www.statista.com/statistics/1103040/cumulative-coronavirus-covid19-cases-number-worldwide-by-day/
2. https://www.seruminstitute.com/product_covishield.php
3. https://www.bharatbiotech.com/covaxin.html
4. https://www.cowin.gov.in/
5. https://dbtindia.gov.in/
6. https://dbtindia.gov.in/sites/default/files/uploadfiles/Website%20Content%20for%20Mission%20Suraksha-V3.pdf
7. https://main.icmr.nic.in/sites/default/files/annual_repoorts/ICMR_AR_English_2020_21_final.pdf
8. https://cdsco.gov.in/opencms/opencms/system/modules/CDSCO.WEB/elements/download_file_division.jsp?num_id=ODE0MQ==
9. https://esanjeevani.mohfw.gov.in/assets/guidelines/Telemedicine_Practice_Guidelines.pdf
10. https://nidhi.dst.gov.in/cawach/
11. https://isba.in/cawach/
12. https://www.meity.gov.in/writereaddata/files/Innovation_Challenge_VC.pdf
13. https://ikpknowledgepark.com/
14. https://www.indiascienceandtechnology.gov.in/covidfunding/ico-fund-ikp-covid-19-fund
15. https://niscpr.res.in/includes/images/annualreport/annual2021-23.pdf

19 Digital India: A Boon Harnessed Well—Co-WIN—Enabling over 2 Billion Vaccines

R. S. Sharma

19.1 Introduction

The development of the COVID-19 vaccine has been one of the most amazing feats of human engineering and coordination under pressing time constraints. With the discovery, trials, and approvals behind us, the next step in this large-scale immunization program was to produce, procure, and distribute the billions of vaccines that would be needed for regular life to resume. Moreover, we had to do this with extreme speed. If the vaccine program was not fast, we would have been left chasing our own tail, as those vaccinated early may have lost immunity before the last person was vaccinated.

A successful drive was contingent upon smooth coordination between a large number of stakeholders, requiring extreme speed from all of our health infrastructure, both private and public. With uncertainties and anxieties in full swing, conditions were also ripe for rent-seeking, black-marketing, and other ugly unfairness to prop. While no system can guarantee that there is no fraud, we still needed to install checks and balances to minimize it. The biggest check was transparency around the vaccination data.

Additionally, there were various considerations to keep in mind as vaccinations would have ensued—citizens' preferences in the type of vaccine, timing of vaccination, and place of vaccination. In parallel, vaccination centers needed flexibility and agency too, to decide how best to serve their local constituencies.

In the COVID-19 vaccination scenario, extreme speed, smooth coordination, and vaccine efficiency were paramount. We needed to balance these three while also triaging fairly, providing individual agency and flexibility to vaccination centers.

R. S. Sharma (✉)
National Health Authority (NHA) of India and Chairmen, Empowered Group of COVID Vaccination (EGVAC), New Delhi, India
e-mail: rssharma3@gmail.com

© The Editor(s) (if applicable) and The Author(s), under exclusive license to Springer Nature Singapore Pte Ltd. 2025

R. Swarup et al. (eds.), *At War with the Single Strand*,
https://doi.org/10.1007/978-981-96-8122-8_19

This was hard for any country, but an even harder challenge given our population size, with the diversity and federated structure of India.

We needed an information system that acted as a single source of truth to ease coordination between the actors, enforce citizens' agency, and prevent unfairness stemming from information asymmetry. We also had to be aware of the reality that not all states would have had the same capacity or spending power to develop such a system. If an individual had taken a vaccine dose in one state, multiple siloed systems (or lack of a system) should not have been the reason why they were denied their second vaccine dose in another state.

To reopen socio-economic activities, many sensitive locations, and international borders eventually required vaccine certificates to safeguard the interests of their respective residents. This meant individuals needed provable ways of recording the date, time, and brand of vaccine administered to them. The final crinkle was that this was not a use-and-throw-away system. The information infrastructure we built needed to be able to serve citizens in the long run, as at the time, it was not known whether COVID-19 vaccinations were here to become an ongoing program.

This is where Co-WIN stepped into the picture to answer the multitude of worries raised in the points aforementioned. Co-WIN served as the digital information management system that bridged the gaps to enable the administration over 2 billion vaccine doses in a record time.

19.2 About Co-WIN

Co-WIN or the Covid Vaccine Intelligence Network served as the technological backbone of India's mass vaccination campaign. The Empowered Group on Vaccine Administration was the committee driving the development on Co-WIN. The production of Co-WIN was jointly overlooked by the Ministry of Health and Family Welfare (MoHFW) and Ministry of Electronics and Information Technology (MeitY).

Demonstrating India's IT prowess, backed by its Digital Public Infrastructure, the platform was perfected over just 2 weeks and was ready in time to roll out in January 2021 as soon as Emergency Use Authorization was given to the two vaccines. The indigenously developed technology was launched along with India's vaccination program on January 16, 2021 by the Hon'ble Prime Minister Shri Narendra Modi.

Co-WIN served as a single source of truth for all stakeholders. Formulated as a transparent system, it allowed tracking of each dose of vaccine supplied to vaccination facilities and recording the fulfillment of demand at a granular level. Co-WIN provided a personal dashboard to vaccinators to view the performance at their facilities. We needed thousands of healthcare centers to manage logistics and inventory while being vaccine-efficient. The Co-WIN data platform made their jobs easier. Further, the data collected was made available on an online dashboard on Co-WIN's website to administrative authorities to view key indicators for vaccination coverage. This helped administrators answer important questions about the demand for vaccines in their constituency, as well as the performance of vaccination centers.

More importantly, this same data was made available to the larger public as well. This kind of transparency is what kept a check on rent-seeking and made distribution equitable at all nodal points. This degree of traceability removed information asymmetry from the entire ecosystem and made all stakeholders accountable.

In Fig. 19.1, we see how Co-WIN streamlined the flow of information for vaccine delivery from the cold-chain points down to the last resident of this country.

One of the more critical roles for an information management system like Co-WIN was also to act as a *feedback* tool for India's vaccination policy. Co-WIN enabled data-driven public health policy and helped evaluate the performance of each brand of vaccination. The Co-WIN platform had been used to learn from the frontlines and standardize the quality of vaccine delivery across vaccination facilities. Additionally, Co-WIN also had the capability of recording the After Effects Following Immunization (AEFI) data. While we are extremely lucky that the current set of vaccines have shown next to no adverse effects in the population, let us remember that the unfortunate circumstances had not provided our scientists and researchers with sufficient time for multiple testing cycles for these vaccines. We could just have easily had problems in the roll-out that were not caught in the trials. An AEFI data system served as our early-warning system to pause or in extreme cases even terminate the roll-out if a vaccine were found to introduce unacceptable side effects.

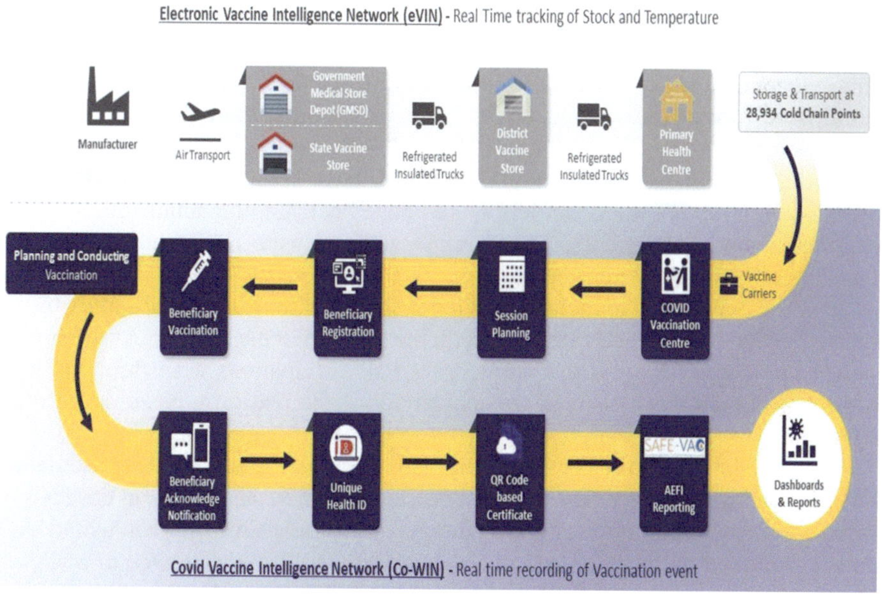

Fig. 19.1 Supply chain of COVID-19 vaccines

With the evolving nature of the pandemic, it was expected that any system designed for vaccine administration would have to be dynamic and adaptable to accommodate new changes. This had been evident from the different phases of vaccination roll-out that had seen evolving policies on vaccine administration and growing needs of citizens and vaccinators using the platform. We also had to be prepared for the unexpected. In anticipation of parallel systems taking place, the entire system was designed as a *digital public good* based on open APIs (Application Programming Interfaces), with due security measures in place. These open APIs had been made public to further ensure innovation on top of the Co-WIN platform to drive wider coverage and inclusion by third-party developers.

To ensure the system did not become a victim of a digital divide, various measures were taken to ensure Co-WIN served the length and breadth of this nation equitably. Figure 19.2 shows a summary of the features offered for citizens and vaccinators alike by Co-WIN.

In Fig. 19.3, we further explore how Co-WIN ensured inclusivity for a nation as diverse and regionally different as India. Co-WIN provided multiple modes of registration in walk-ins, online, and assisted registrations through Community Service Centers (CSCs) and call centers. Second, digital self-registration had been simplified and made available in multiple languages. Third, minimal data inputs were required, and multiple options of photo-identity proofs had been provided to smoothen the process of vaccination.

There was, however, a unique challenge that the COVID-19 virus posed. Crowds of citizens overwhelming the vaccination centers was a risk. The third phase of the vaccine was opened to 865 million citizens at once. When the registration for 18–44 age group was opened, if there was no system of presence-less allotment, we would have witnessed chaos at the centers. Not only would this have endangered the crowd but also the frontline healthcare workers. Such a chaos could have further created incentives for people to distribute fake vaccines and start black-marketing and profiteering from the access to vaccines. There was also a risk that tension at a vaccination center may have spilled over into a law-and-order issue. Even one such incident would have been amplified and added to vaccine hesitancy in the public. This is why walk-ins were not opened up for the 18–44 age group at the very beginning.

Hence, Co-WIN had to include a digital system of allotment till the supply by the vaccine manufacturers had caught up with the demand in the country. As the supply caught up, it was observed that over 80% of all vaccinations were through direct walk-ins. This further bolstered the inclusivity of the platform, ensuring no digital divide.

Flexibility also introduced an additional challenge: tracking the vaccination schedule and the brand of first dose of vaccine and when the follow-up dose has to be administered. Both for the digitally literate and illiterate, the problem would have gotten amplified as more brands of vaccines entered India. *Issuance of a digital certificate* to record the vaccination helped guide all individuals when they visit for their second dose.

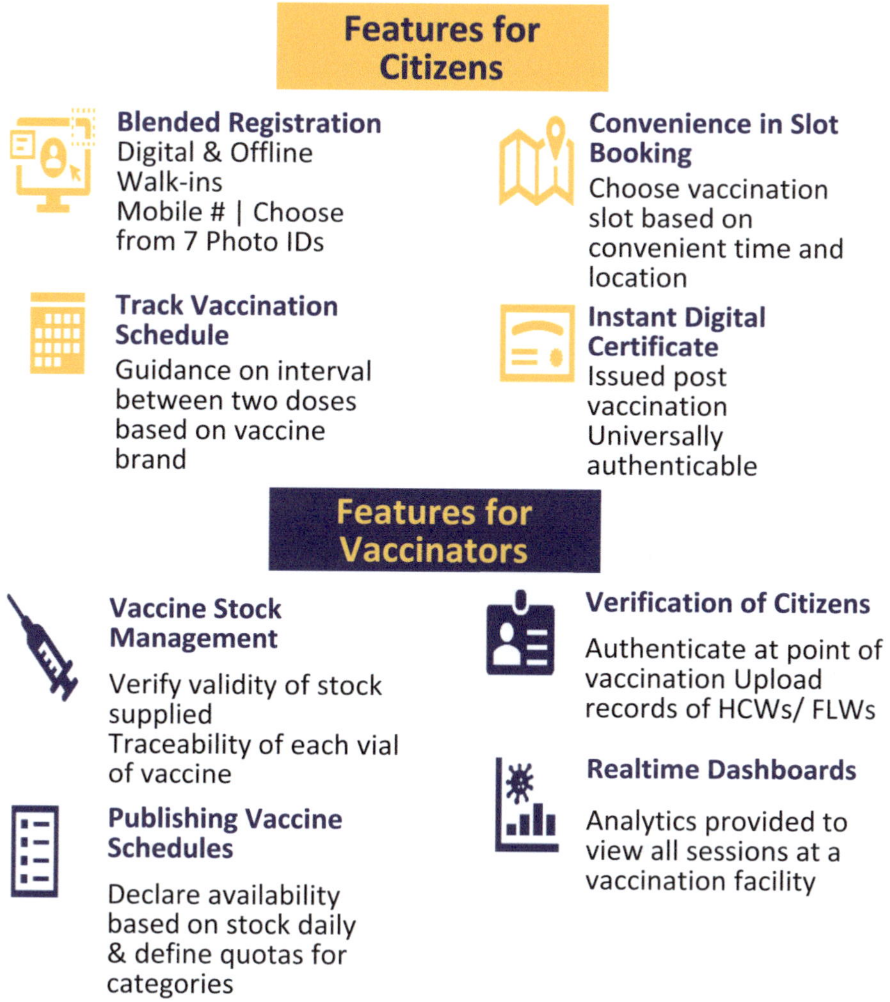

Fig. 19.2 Co-WIN features for citizens

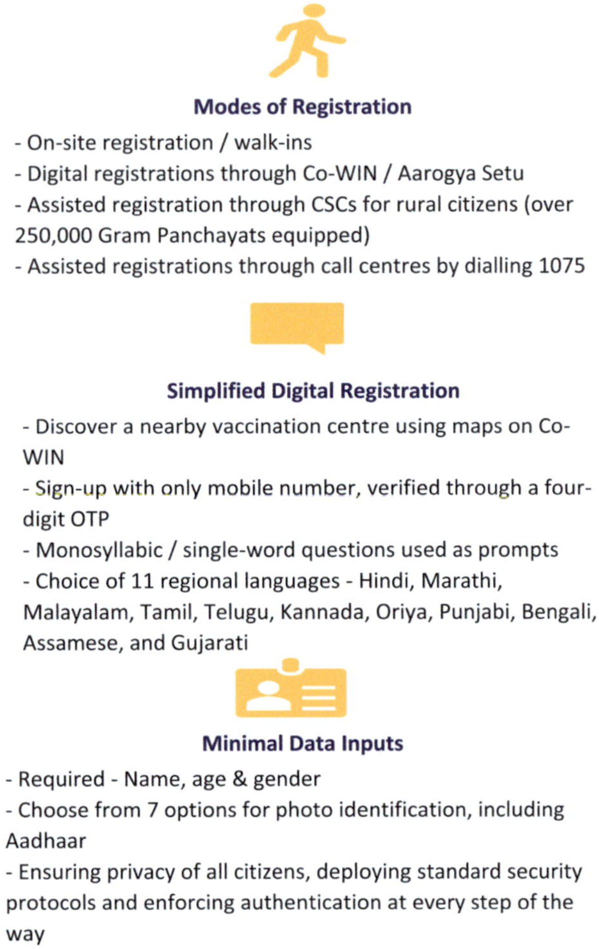

Fig. 19.3 Co-WIN registration

19.3 Evolution of the Co-WIN Platform

In Phase I starting January 16, 2021, Co-WIN focused on assisting healthcare workers (HCWs) and frontline workers (FLWs). All states and union territories had to create and share lists of HCWs and FLWs to be uploaded in bulk. At this point of time, there was no online appointment system. The government rolled out vaccination by prioritizing the most vulnerable.

HCWs and FLWs were slotted for vaccination through Co-WIN at specific facilities, with text messages sent with the date and locations of vaccination. However, the first challenge was unclean data and incorrect phone numbers. As a result, individuals were not receiving text messages. Furthermore, due to the nature of their jobs, at times, HCWs and FLWs were not able to attend their vaccination slots as they were required to travel to places they had not opted for. This supply-driven approach resulted in the lower efficiency of vaccination centers, resulting in the underutilization of manpower. Co-WIN data helped see that this approach would not work, especially when it would be applied to the general public at large. This resulted in the first fundamental change in the vaccination policy—demand-driven distribution. The government can make vaccines available, but individuals have to choose to take them.

19.3.1 First Fundamental Change: Approach Should Be Demand-Driven

Further, it was initially decided that the second dose of the vaccine would be administered at the same center to ensure that the individual gets the same vaccine. On deliberation, we realized restricting citizens based on location for the second dose would reduce compliance and increase complications in the administration. Hence, the second fundamental change was brought about to the vaccination policy.

19.3.2 Second Fundamental Change: Flexibility in Booking Second Appointment at Any Centre

The guiding principle of ensuring that the system can adapt quickly based on the rapidly changing socio-economic and policy environment was always kept in mind. An *open architecture with interoperability* was the result of this driving principle. It was anticipated that large hospital networks may choose to use their IT systems for vaccination, and we are now seeing states produce their native systems. The open APIs help us to ensure that all systems are talking to each other and there remains *a single source of truth.*

The Phase II of vaccine roll-out saw the target audience doubling. Vaccinations were opened for citizens aged 60 and above and 45 and above with comorbidities. This was followed by vaccination opening for all citizens aged 45 and above. With a larger audience, the online self-registration and appointment system was launched for the first time. Changes were also incorporated in the vaccinator modules to accommodate larger crowds, differentiation based on age. Further, capabilities of data reporting and analysis were also added, along with a revamped view for dashboards.

Phase III was the most challenging phase of the development. With vaccination being opened to all citizens above the age of 18 (865 million in total), online registrations became a necessity for managing demand at vaccination centers. More importantly, to handle nearly a billion visits/visitors on the portal, significant developments were undertaken for capacity building to make the backend of the system robust. Capabilities to handle different vaccination policies were based on state governments. To ensure inclusivity, Co-WIN was also adapted for lingual diversity and disability-friendliness.

19.4 Architecture and Interoperability of the Co-WIN Platform

The evolution of Co-WIN has shown how it has served as a much-needed feedback mechanism to help course-correct our vaccine policy. It also helps enforce a complex policy in a simplified way across hundreds of thousands of stakeholders in the vaccine distribution supply chain. While we have been lucky with AEFI numbers, Co-WIN would have helped give early signs of those too.

The engineering and architecture of the Co-WIN platform also deserves its due credit, which allowed for such swift and fundamental changes to both the vaccine policy and citizen access. Figure 19.4 lays out the architectural diagram of the Co-WIN platform which explains its modular construct that allows for interoperability.

19.4.1 Architecture of the Co-WIN Platform

The Co-WIN platform has four modules: orchestration module, citizen registration module, facility management and vaccinator module, and the certificate, feedback, and AEFI reporting module.

Each of these four modules functions independently from the others, which minimizes dependencies across modules and makes it easier to fix bugs or errors. At the same time, each module feeds information into the next one, which minimizes user labor and redundancies.

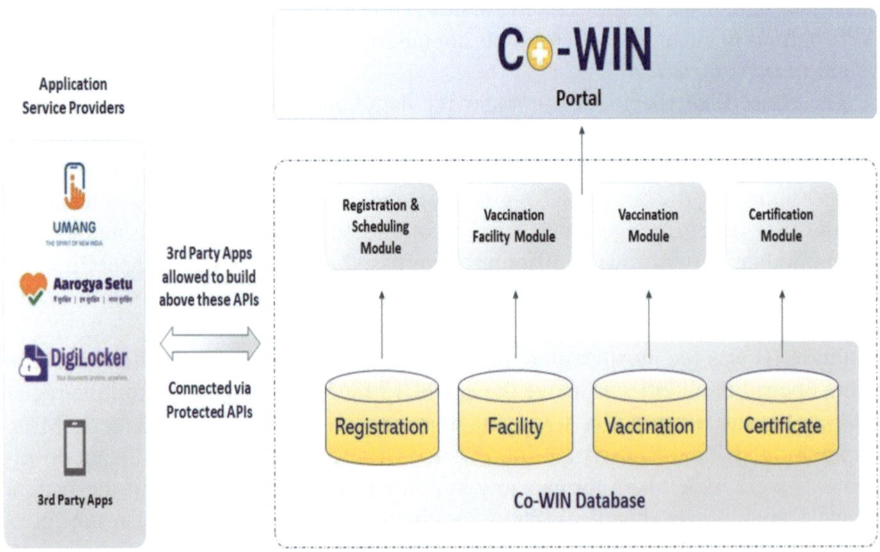

Fig. 19.4 Architecture of the Co-WIN platform

Through the first or the orchestration module, Co-WIN administrators, at the national, state, and district levels, are registered on the platform using unique logins. Administrators in turn assign roles to other system users, such as vaccinators and verifiers. Using the databases of vaccine stocks and sites, state authorities supply vaccines to the districts, and districts supply them to the vaccine facilities. District administrators can acknowledge the receipt of vaccine stocks by uploading the batch number of the stocks received through the application.

The second module of citizen registration allows citizens to register themselves for the vaccination program, either through the Co-WIN website or through other government and private applications such as Aarogya Setu and UMANG. Registered citizens can select their preferred date, time slot, and vaccination center. To circumvent the lack of digital access, the platform allowed four members to be registered using one user login, which was eventually extended to six once vaccinations for children were rolled out. Additionally, there is provision for people to walk into vaccination centers, where workers can register them on the platform. Registered users receive automated text messages on their registered mobile numbers regarding vaccination appointments, second dose reminders, and links to digital vaccination certificates.

The third module is the facility management and vaccinator module which is used by vaccination facilities and officers, verifiers, and supervisors at the vaccination centers to update the status of vaccinations in the backend. Vaccination facilities would update the schedules on a daily basis on the number of slots available for vaccinations based on the stock present at the facility. Additionally, sessions could be managed and tracked on a live basis using this module. Vaccinators would also record the event of vaccination using this module, after verifying the identity of the resident getting vaccinated.

Lastly, the fourth module, certificate, feedback, and AEFI module, helps administrators issue proof of vaccination. The module also provides a feedback mechanism for vaccinated people to report and fix errors. It also enables reporting of adverse events following the vaccination using SAFEVAC, a surveillance system integrated with Co-WIN.

As an additional measure of ensuring accessibility and inclusivity for all, the interoperability enabled by Co-WIN's architecture played an instrumental role. With access to different modules made available through protected and public APIs, various public and private organizations were able to join hands with the government to leverage their existing IT infrastructure to contribute to the national vaccination campaign. Figure 19.5 illustrates an example of how different organizations and applications could come together to carry out an entire cycle of vaccination.

The partnerships for various oragnizations were carefully monitored and processed through Co-WIN'S sandbox environment. Over 300 applications were received, out of which over 150 players completed integrations successfully.

Depending on the nature of operation of an organization's business, each entity integrated with different or all APIs made available. For instance, various discovery platforms and health-tech applications integrated with the citizen registration APIs to provide more fronts for citizens to view and book their vaccinations. State governments and hospital chains utilized the vaccine facility APIs, and travel and communication applications integrated with certification APIs.

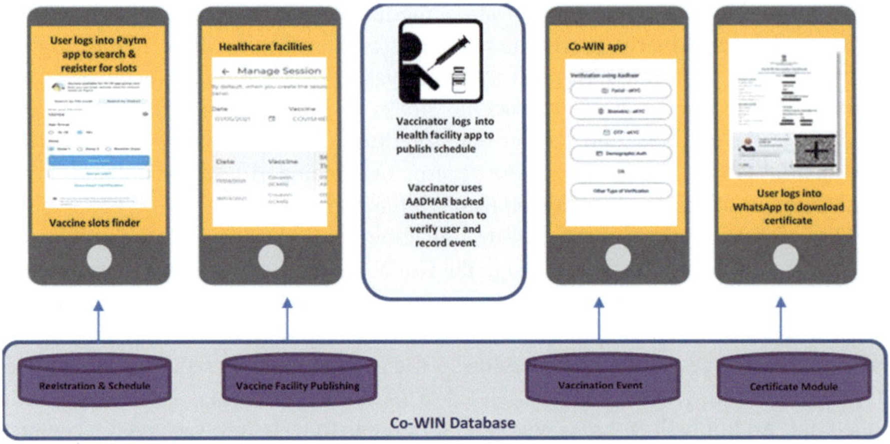

Fig. 19.5 Illustration of interoperability enabled by Co-WIN

19.5 Co-WIN'S Impact

As of 2022, Co-WIN has over 1.1 billion citizens registered on it through various media. Over 2.2 billion events of vaccinations were recorded on the platform, without a single day of downtime. In the world of digital technologies, this is no ordinary feat. Figure 19.6 illustrates a view of the public dashboard made available to all stakeholders in the country.

This public dashboard enabled authorities and individuals to granularly monitor the coverage of India's COVID-19 vaccination campaign. With drill-downs available down to the district level, all stakeholders could evaluate the rate and coverage of vaccinations in near real time. The dashboard also provided the capabilities to slice and dice the data based on a number of parameters, which included splits between the first and subsequent doses, by age and gender and by vaccine brands. This degree of transparency enabled by Co-WIN is what allowed for an equitable distribution and no rent-seeking behavior to prop up for even a single day.

The strength of a technical system to sustain heavy traffic is often overlooked by the want of features and glamor of UI/UX. For a system that cannot afford to fail as real lives were at stake, Co-WIN's ability to sustain a traffic of hundreds of millions of users on a daily basis was truly commendable. For instance, when the portal was opened for the age bracket of 18–44 years, the platform was able to handle 13.7 million registrations that translated in over 3 billion visits on the website, within 8 h of being launched, as illustrated in Fig. 19.7.

If evaluated from the lens of existing mass-scale tech platforms, Co-WIN would stand out among all its peers who have the same volume of users. Co-WIN broke all records when it came to scaling every milestone of 100 million users. From registering 200 million users in merely 4 months to over 1 billion users in just 10 months, Co-WIN's robust engineering always marched from strength to strength.

Fig. 19.6 Co-WIN dashboard. (Source: https://dashboard.cowin.gov.in/)

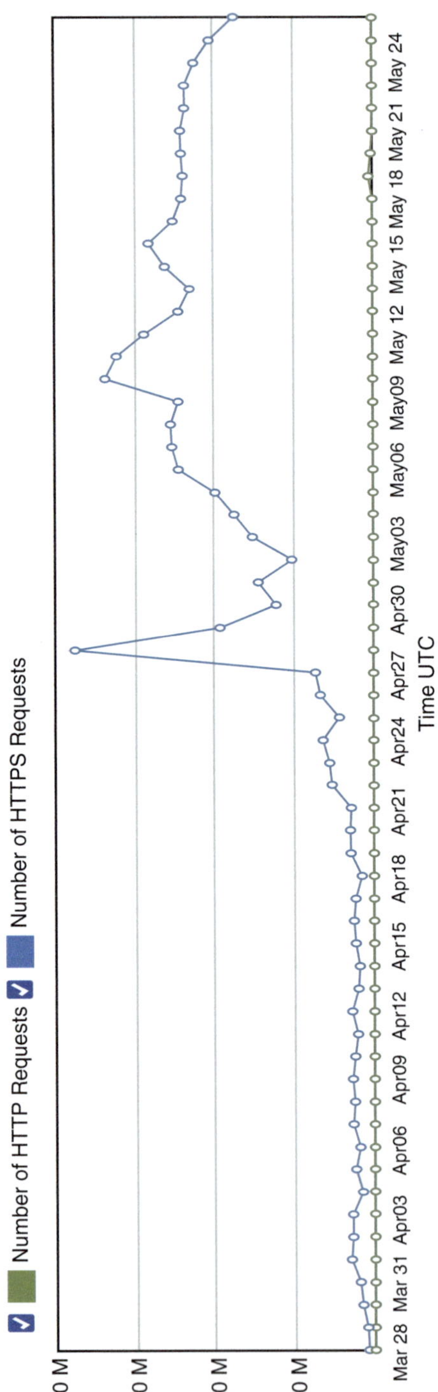

Fig. 19.7 Internet traffic observed on Co-WIN post-launch of the public registration portal

The biggest feat achieved by the platform was on September 17, 2021, when it enabled the recording of 25 million vaccinations in a single day. While the true credit for this achievement goes to our healthcare and frontline workers (HCWs and FLWs), Co-WIN was able to withstand nearly 1% of all the Internet traffic in India. This translated into 65,000 API hits per second. Such a high hit rate could have sent the best of systems into a disarray, but the technological prowess of Co-WIN ensured that the efforts of our HCWs and FLWs in achieving the world record of vaccinations in a single day was not hindered for even a single second.

19.6 Co-WIN'S Global Acceptance and Beyond

The certificate module of Co-WIN was based on the indigenously developed verifiable credential—DIVOC (Digital Infrastructure for Verifiable Open Credentialing). The use of this cutting-edge technology in each of the vaccination certification ensured that not a single vaccination certificate issued could be tampered with.

As we all fought with the pandemic, this module of Co-WIN ensured that as socio-economic activities were opened up in a phased manner, the residents of India could travel both domestically and internationally without any hesitation. With mutual recognition with over 130 countries, Co-WIN had received global acceptance, which allowed for the smooth travel of Indian residents.

The first successful use case of verifiable credentials at a population scale was made possible through Co-WIN. It ensured that we have received interest from 76 countries thus far. A total of 196 foreign officials and 116 individuals from 41 countries have registered for participation in the Global Conclave on Co-Win, scheduled for July 5, 2021. In due course of time, we will also have information on the countries that formally showcase their interest in adopting the platform for vaccination in their respective countries.

Considering the success and maturity of the platform, the Government of India also made the technology of the Co-WIN platform open source to all interested governments and countries globally for free, till perpetuity. Since Co-WIN is a public digital good in India, it was intended to be shared with other countries with the same expectation. For the same, the Co-WIN Global Conclave was conducted on July 7, 2021, which witnessed participation from over 141 countries and UN representatives.

Competing Interest The authors have no conflicts of interest to declare that are relevant to the content of this chapter.

COVID 19: The Global Response and Our Learnings: Lessons from the Pandemic for Science and the Way Forward

20

Soumya Swaminathan

The unprecedented challenges which were faced by all countries, including India, clearly highlighted the urgency to invest in health to maintain the economic well-being of society. COVID also showed us that rapid advances in science and technology and the better understanding of the virus and its pathogenesis helped us to use that knowledge to develop the tools to conquer the virus. The focus clearly was on investments in public health, especially primary health care and health of our work-force. The need for enhanced investments for essential commodities like oxygen, ventilators, PPEs, etc. was also critical. The Covid pandemic also exposed the lack of global solidarity and the role of geopolitics in responding to a global threat, with nationalism and shortsighted policies leading to vast inequities in the access to countermeasures across populations. It also exacerbated the existing inequalities in society, so that the most vulnerable and marginalized in every country suffered the worst consequences [1]. Further, people with underlying co-morbidities and non-communicable diseases developed more severe disease after infection. We also observed a new phenomenon of "infodemic", where people were exposed to huge amounts of information that included a lot of mis- and disinformation. This confused people and contributed to vaccine hesitancy and loss of trust in authorities.

Noncommunicable diseases or NCDs like heart disease, cancer, diabetes, chronic respiratory problems, stroke, etc. are a major cause of premature mortality For effective prevention of NCDs, we need special attention on many risk factors like our lifestyle, physical activity, proper diet, and other factors like consumption of alcohol, tobacco, and environmental pollution. For the management of these factors and prevention and control, a whole of Government approach with the participation of private sector, civil society, and other stakeholders is essential. We need effective policy actions to tackle the issue of commercial determinants of health (which can be positive or negative)—good

S. Swaminathan (✉)
MS Swaminathan Research Foundation, Chennai, India
e-mail: doctorsoumya@yahoo.com

© The Editor(s) (if applicable) and The Author(s), under exclusive license to Springer Nature Singapore Pte Ltd. 2025
R. Swarup et al. (eds.), *At War with the Single Strand*,
https://doi.org/10.1007/978-981-96-8122-8_20

examples are the banning of transfats in food, tobacco regulation and taxes, incentivizing the consumption of millets by making them available through the PDS, and regulating the sale of high-fat, high-salt, and high-sugar food and beverages in places near and around schools. Hence, both regulation and the use of fiscal tools (taxes) can play an important role in incentivizing positive behaviors and reducing risk factors in the community. Such actions will help to create a healthier population, with a better quality of life [2].

To overcome the acute challenge of the pandemic, there was a global effort for the development of a large number of safe and effective vaccines. These were developed in a very unprecedented time frame of less than a year and administered successfully across the globe. More than 14 billion doses were administered in 2 years of the virus infection. The Indian Vaccine Industry played a very important role of developing indigenous vaccines not just for the Indian population but also supplied to other countries through Covax and directly. However, it is estimated that over a million lives could have been saved in 2021 globally if we had equitable access to the vaccines. To ensure equitable access to life-saving products like vaccines and drugs during a global health emergency, special urgent attention is needed for intellectual property management, technology transfer, and assured quality manufacturing through a distributed manufacturing network, including on the African continent. The pandemic accord being negotiated by WHO member states is a once-in-a-lifetime opportunity to address this issue, so that the same mistakes are not repeated [3].

India did well in terms of innovation, R&D, and its massive development and scaling manufacturing of diagnostics and vaccines, apart from oxygen, ventilators, etc. While there were initial hiccups in supply chains and lack of PPE etc., these were quickly overcome by the government and manufacturers working together. A lesson to be learnt for the future is to have a good network of hospitals and clinical trial centers that can rapidly undertake research studies as needed, including clinical trials of new drugs and vaccines, or other treatment approaches. This requires advance planning, ready-to-use protocols, ethics, and regulatory approvals in real time and an efficient logistics system in place. The UK was able to mount large clinical trials using their NHS network of hospitals, with patient data digitally collected and managed centrally. The result was that in July 2020, they found that steroids reduce mortality due to Covid in hospitalized patients, and this quickly became the standard of care globally [4].

To prepare for future outbreaks of known and emerging pathogens, India needs to invest in good surveillance, data capture, management, analysis, and response. Multidisciplinary teams are needed at the district level (ideally) who can collect data from routine or sentinel surveillance and are trained to pick up unusual signals for further investigation. It is preferable to undertake One Health surveillance, which encompasses human, animal, and environmental surveillance. Genomic technologies can now be applied to wastewater to monitor the presence or load of various bacteria, viruses, and other pathogens [5]. This has been used successfully by some cities to warn them of impending Covid waves and has been employed for a long time to monitor poliovirus circulation in communities.

Many new technologies were developed and used in the past few years. Some can be modified or redesigned to tackle our other major public health problems. For example, there is a great need for rapid, point-of-care diagnostics for infectious diseases like dengue, scrub typhus, and tuberculosis. We need better vaccines against TB and malaria. And we also need better treatments for cancer and neurodegenerative diseases that are increasing rapidly. The platform technologies that were developed should now be used. For example, mRNA technology has huge potential not only for the development of novel vaccines against many pathogens but also for the development of monoclonal antibodies against cancer and immune system disorders. With the government prioritizing and funding early-stage research into these (and other novel) technologies, it is more likely that the private sector will come forward to work on these problems. In the USA, for example, BARDA and DARPA, both of which have a national security goal, have funded the development of many innovative health products in the past decades. Health security is an important component of national security and should be given high priority.

Attention also needs to be paid to the regulatory system (CDSCO) for drugs, diagnostics, and vaccines. A pathway needs to be created to fast-track products that are of high public health priority, without compromising on the quality of the assessment for safety and efficacy. New therapies including gene therapies and biologics are going to be more common, and capacity has to be built to assess them properly. The regulatory agency needs to promote innovation while safeguarding the public health of citizens. Raising the profile of the drug regulator (DCGI) to have an independent, highly competent, and well-respected expert at the level of Secretary to the Government of India (like the FDA Commissioner in the US) would help bring the necessary gravitas and competence that are required of this agency.

India was quickly able to roll out applications for contact tracing, vaccination, etc. as well as telemedicine services because of the attention given to digital infrastructure, with the national digital health mission defining the standards as well as the governance and vision. We have excellent specialists in our cities who could now also apply their skills in remote rural populations and thereby improve the quality of healthcare. The digital health services should be expanded and strengthened to leverage this capacity.

Investments in primary health care, including regular training and mentoring of the workforce and use of digital tools to support them are going to be important. The need for a separate public health cadre in each state and at the center cannot be overemphasized. We also need to consider the creation of a separate cadre of community-level health officers who can manage PHCs (like physician assistants) [6]. The difference between clinicians who treat sick people and a public health cadre (including doctors, epidemiologists, nurses, nutritionists, entomologists, sanitary inspectors, and laboratory technicians) who work on disease prevention and health promotion must be recognized and appreciated.

Science and health literacy are going to be key factors for an enlightened and healthy population. This must start in schools, but people of all ages need credible and high-quality information presented to them in easily understandable ways [7]. Attention to ventilation is important for all airborne respiratory pathogens. Either good natural ventilation or upgraded HVAC systems should be in place in all settings where a lot of people spend time indoors (educational institutions, workplaces, malls, restaurants, etc.). Maintaining good health often requires behavior change, social support, and incentives; our public health cadres must include social scientists, behavioral economists, and anthropologists, and communication strategies need to be tailored to social and cultural contexts.

The pandemic has taught us many lessons, and we need to learn them to be better prepared next time. The issue of another pandemic is not if but when it will occur. We need to take a holistic view of health management, which addresses the needs of sick persons coming to our health facilities, and also focus on the prevention of diseases and promoting health wellness—both physical and mental. The world is facing many challenges including the impacts of climate change on health, food systems, and biodiversity, related environmental risks like air pollution and water scarcity, as well as the increasing threat of antimicrobial resistance. Only if we have a strong and resilient health system, as well as a robust and responsive R&D ecosystem, can we protect our 1.4 billion Indian citizens against various health threats [8].

Competing Interest The authors have no conflicts of interest to declare that are relevant to the content of this chapter.

References

1. World Bank. World Development Report 2022: Finance for an Equitable Recovery. Washington, DC: World Bank; 2022. https://doi.org/10.1596/978-1-4648-1730-4.
2. 2021 Global Nutrition Report: The state of global nutrition. Bristol, UK: Development Initiatives, 2021.
3. Progress and challenges with achieving universal immunization coverage. WHO/UNICEF Estimates of National Immunization Coverage July 2022.
4. https://www.who.int/teams/blueprint/covid-19
5. https://www.who.int/publications/i/item/9789240052857
6. Brauner JM, et al. Inferring the effectiveness of government interventions against COVID-19. Science. 2021;371, eabd9338 https://doi.org/10.1126/science.abd9338.
7. Norton A, Bucher A, Antonio E, et al. A living mapping review for COVID-19 funded research projects: one year update [version 5; peer review: 2 approved]. Welcome Open Res. 2022;5:209.
8. https://ourworldindata.org/covid-vaccinations

Epilogue: The Road Map Ahead—Preparedness for the Future

Renu Swarup, Narendra Kumar Arora, and G. Padmanaban

The COVID pandemic has been etched in history as a public health crisis which impacted the health, economy, and the social condition of the society. A pandemic which was witnessed after over a century—the Spanish Flu Pandemic—had struck in 1918 and lasted over 2 years following the World War. COVID devastated the world. As per WHO statistics, a total of nearly 775 million cases were infected and over 7 million deaths reported worldwide since the first case was reported from China in December 2019. The disease was identified as an infectious respiratory disease caused by an RNA virus from the Coronavirus family. WHO gave it an interim name—the novel coronavirus-2019—nCov. On February 11, 2020, WHO gave it an official name COVID-19, and the disease was SARS-CoV-2. The disease outbreak spread at an alarming speed, and on March 11, 2020, it was officially declared as a pandemic by WHO. In India, the first infection was reported in January 2020. As the disease progressed, the virus variants caused different waves—the first being the variant of concern, Alpha, which was reported from UK in December 2020, followed by Delta variant reported from India in March 2021. This was one of the most serious waves of infection in India, and this was followed by Omicron. The country saw a comprehensive response with all stakeholders coming together with a collective responsibility to fight this war. It was for the first time that it was a war being fought by each citizen. It was a clear message that conveyed that the world is safe only when each citizen is safe.

Covid was a Public Health Emergency which changed the world not just in terms of our prioritization of public health research but also in terms of our collective

R. Swarup
Former Secretary to Government of India, Department of Biotechnology, Ministry of Science &Technology, New Delhi, India

N. K. Arora
Chairman, Executive Director, INCLEN Trust International, New Delhi, India

G. Padmanaban
Former Director, Hon Professor, NASI Hon Senior, Scientist, Department of Biochemistry, Indian Institute of Science, Bangalore, India

© The Editor(s) (if applicable) and The Author(s), under exclusive license to Springer Nature Singapore Pte Ltd. 2025

R. Swarup et al. (eds.), *At War with the Single Strand*,
https://doi.org/10.1007/978-981-96-8122-8

thinking on the road map ahead. The pandemic highlighted the need for nations to be prepared to respond to health emergencies.

The scientific journey across these phases, with technology, tools, and platforms responding to the advancing disease, has been elucidated in the chapters in Section II. The challenges posed by the COVID-19 pandemic and the role science played in responding to these challenges have been highlighted in the different chapters. The pandemic tested our resilience, and we witnessed how science and technology took giant strides to help India as well as the global community to mitigate the impact of the virus. India was at the forefront of the global fight against COVID-19 to deliver innovative solutions at scale not just for itself but also for the world. The learnings and successes, such as ramping up indigenous testing capacities and collaborating across sectors and communities for innovation and development of indigenous vaccines, and genomic surveillance, have helped us build a stronger ecosystem to fight the health crisis. We are now confident of having a much stronger scientific basis of response to any challenge we may face in the future.

For a successful research ecosystem to deliver, there are three important pillars—capacity, cutting-edge technologies, and collaboration. Our country has seen a special focus on all these three during the pandemic. It is, however, important that this robust ecosystem which has delivered so effectively in this crisis continues to do so for future challenges. Capacities have been built both in terms of human resource and infrastructure not just to respond to the pandemic but also for handling cutting-edge newer technologies which make us future-ready for tackling many other diseases and healthcare challenges. The ecosystem today allows for a very effective collaboration between researchers from academia and industry, the start-ups, and international stakeholders. A state-of-the-art network of shared infrastructure has been created across the product development value chain for all biopharma products, which served as the foundation of our success in the pandemic. SARS-CoV-2 is not the last disease we are fighting a war against, there are, and will continue to be, a large number of pathogens and other health priorities for which there will be a continued focus on science-based solutions to address the challenges of human health. This ecosystem which we have built was put to test during the pandemic with high levels of success, and this strong ecosystem is now ready to deliver.

COVID gave us the understanding of the importance of investing in R&D, the importance of having well laid out systems for research and translation, and accelerated pathways to product development and also highlighted the strength of global collaboration and knowledge-, data-, and resource-sharing. COVID-19 resulted in large amounts of research funding being mobilized during the pandemic. But there is still limited R&D preparedness capacity, specially in terms of research capacity for human resource and infrastructure, technology transfer, and research resources including clinical trial sites.

To make the best use of limited funding, we need enhanced international coordination for R&D. Although we have seen increased coordination of R&D during the pandemic, the R&D ecosystem still needs further strengthening to rapidly switch into emergency mode when needed. We worked collectively together to develop medical countermeasures rapidly and developed accelerated pathways for diagnostics and vaccine development and manufacture including a rapid response regulatory approval framework; however, we need similar mechanisms for rapid therapeutics development. To be prepared for future threats, we need to focus on priority pathogens and develop institutionalized mechanisms which encourage and promote high-risk research and product development.

The focus on a strong surveillance system is essential not just to have an early identification of the pathogens, but through a proper correlation of epidemiological data with clinical data, appropriate disease management and treatment strategies can be developed. Regional surveillance networks are also essential to monitor the transboundary movement of pathogens, especially considering the One Health approach. Globally accessible repositories are important for resource and data sharing which will help plan both new surveillance and monitoring strategies and also the development of new products.

It is also essential to continue to encourage the important academia industry partnerships which were critical in our fight against the pandemic. It is also important to focus on global collaboration models and develop a framework which allows innovations to move across borders and boundaries

We need to now take forward our learnings from COVID.

- Accelerate progress on emerging technologies
- Collaborate for more effective action
- Invest for a sustainable response
- Innovate to secure the future

This book has attempted to bring to the readers the story of how science fought this pandemic war, in the words of the key players who were instrumental in leading the effort. The road map indicated for the future is based on their suggestions, keeping in mind the learnings and the challenges faced.

This is the story of how Science fought the single-strand virus. Let us collectively work so that we can take necessary actions to be future-ready.

Competing Interest The authors have no conflicts of interest to declare that are relevant to the content of this chapter.

> **Nothing in life is to be feared, it is only to be understood. Now is the time to understand more so that we fear less—Marie Curie**

MIX
Papier aus verantwortungsvollen Quellen
Paper from responsible sources
FSC® C105338

If you have any concerns about our products,
you can contact us on
ProductSafety@springernature.com

In case Publisher is established outside the EU,
the EU authorized representative is:
Springer Nature Customer Service Center GmbH
Europaplatz 3, 69115 Heidelberg, Germany

Printed by Libri Plureos GmbH
in Hamburg, Germany